# THE FRONTIERS COLLECTION

# THE FRONTIERS COLLECTION

*Series Editors*
A.C. Elitzur    Z. Merali    T. Padmanabhan    M. Schlosshauer
M.P. Silverman    J.A. Tuszynski    R. Vaas

The books in this collection are devoted to challenging and open problems at the forefront of modern science, including related philosophical debates. In contrast to typical research monographs, however, they strive to present their topics in a manner accessible also to scientifically literate non-specialists wishing to gain insight into the deeper implications and fascinating questions involved. Taken as a whole, the series reflects the need for a fundamental and interdisciplinary approach to modern science. Furthermore, it is intended to encourage active scientists in all areas to ponder over important and perhaps controversial issues beyond their own speciality. Extending from quantum physics and relativity to entropy, consciousness and complex systems—the Frontiers Collection will inspire readers to push back the frontiers of their own knowledge.

More information about this series at http://www.springer.com/series/5342

For a full list of published titles, please see back of book or springer.com/series/5342

Alexander S. Mikhailov · Gerhard Ertl

# Chemical Complexity

Self-Organization Processes in Molecular
Systems

 Springer

Alexander S. Mikhailov
Abteilung Physikalische Chemie
Fritz Haber Institute of the Max Planck
  Society
Berlin
Germany

Gerhard Ertl
Abteilung Physikalische Chemie
Fritz Haber Institute of the Max Planck
  Society
Berlin
Germany

ISSN 1612-3018          ISSN 2197-6619  (electronic)
THE FRONTIERS COLLECTION
ISBN 978-3-319-86147-0          ISBN 978-3-319-57377-9  (eBook)
DOI 10.1007/978-3-319-57377-9

Printed on acid-free paper

This Springer imprint is published by Springer Nature
The registered company is Springer International Publishing AG
The registered company address is: Gewerbestrasse 11, 6330 Cham, Switzerland

# Preface

A concise definition of *complexity* might be: "The whole is more than the sum of its constituents". Usually, chemistry is concerned with the interactions between individual atoms or molecules, and such interactions can lead to the formation of condensed matter with a high degree of the "dead" order at equilibrium. But the experience tells us that, in biological systems, quite different other processes of order formation may take place, prompting in the beginning of the twentieth century even to ask whether new physical laws had to be found in order to explain such "living" order. Since the lectures on "What is Life?" held by E. Schrödinger in 1943, one knows that this is not needed and that self-organization phenomena can also be observed in rather simple inorganic systems with only small reacting molecules if these systems are out of thermal equilibrium.

Our book intends to provide an outline of underlying theoretical concepts and their experimental verification, as they emerged in the middle of the twentieth century and evolved afterwards. In its style, the book can be regarded as a series of essays on selected topics. Their choice is determined by personal preferences of the authors and reflects their research interests. We do not aim to present a systematic introduction and to review the entire discipline. Particularly, the list of literature references is far from being complete. Since our focus is on the concepts, not methods, mathematical aspects are moreover only briefly touched.

Today, the field is in the state of intense research and much attention is paid to it, as evidenced, for instance, by the Nobel Prize in Chemistry of 2016 for studies of molecular machines. Some of the topics are rapidly developing and are vividly discussed. Nonetheless, we tried not to be biased towards them. In our opinion, it became important to look back and to analyze what has been done since E. Schrödinger has posed his question, and whether we already have an ultimate answer to it.

While finishing the book, we want to emphasize how much we owe to discussions and collaborations with our colleagues, and we would like to express our deep gratitude to all of them. The series of conferences on "Engineering of Chemical Complexity", organized by the Berlin Center for Studies of Complex Chemical Systems, has contributed much to the present work.

Berlin, Germany                                                        Alexander S. Mikhailov
February 2017                                                                  Gerhard Ertl

# Contents

# Chapter 1
# From Structure to Function: An Introduction

Atoms and molecules build all matter around us. Through interactions between them, condensed materials—fluids or solids—are formed. Through reactions, molecules can be transformed one into another and new kinds of molecules can be produced. It is therefore quite natural that first the properties of molecular structures and condensed matter have to be investigated and understood. Traditionally, studies of individual molecules, as well as of interactions and reactions between them, were the subject of physical chemistry as a separate scientific discipline. Over the years, an impressive progress has been made in this field. Structures of most molecules, including such macromolecules as proteins or DNA, have been determined and complex reaction mechanisms have been revealed. Nonetheless, it becomes also clear that even such broad structural knowledge does not straightforwardly lead to the understanding of essential processes in the inorganic nature and in biological cells.

In a major international effort, the complete structure of the human genome has been deciphered within the last decades. But, as this was done, it also became evident that understanding of the genetic system is still far from reached. While all cells in an organism contain the same set of genes, their expression is different and the actual difference is determined by patterns of cross-regulation processes in a genetic network. This situation is not a special property of biological systems, as it may seem at the first glance. Through experimental investigations of heterogeneous catalysis, the mechanisms of many surface reactions became understood. However, it became also clear that there is no straightforward connection between such mechanisms and the reaction course. Due to an interplay between elementary reaction steps, complex spatiotemporal patterns of catalytic activity on metal surfaces can develop and they can change sensitively when environmental conditions are modified.

The task of an architect is to design a house as a static structure and, for this, material properties of construction components should be examined and employed. In contrast to this, a mechanical engineer wants to design a functional device, i.e. a dynamical structure where a set of mechanical parts would be interacting in such a way that desired concerted action takes place. In a similar manner, the task of

© Springer International Publishing AG 2017
A.S. Mikhailov and G. Ertl, *Chemical Complexity*, The Frontiers Collection,
DOI 10.1007/978-3-319-57377-9_1

industrial engineers would be to design the entire factory as a manufacturing system that incorporates various machines. These machines, often forming production lines and conveyor belts, need to operate in a coordinated and predictable way. Of course, construction of a house would also involve a sequence of processes and operation steps. However, such processes are only of transient nature; they are terminated when the final structure is raised. But the operation of a manufacturing factory is persistent—certain processes are repeatedly and indefinitely performed, as long as the factory works. It is such persistent coherent operation that defines a manufacturing system or a mechanical machine.

Persistent functioning can also be characteristic for chemical systems, both natural or synthetic, and artificially designed. Obviously, biological cells are the best example of such a natural system. As long as the cell is alive, many interwoven chemical processes run inside it. In molecular biology, a shift from the structure to the function in the research attitude has already taken place. To understand how a cell works, the knowledge of structures of various molecular components is not enough. Dynamic interactions between these components are essential for their specific functions and such interactions need to be understood in detail. Moreover, patterns of collective operation determined by such interactions should be explored.

Comparing the operation of an industrial factory and a living cell, an important difference can also be observed. In a factory, coordination of various manufacturing processes and operations by machines is to a large extent due to the supervision and control by a human manager or, more recently, by a central digital computer. In the cells, such central control is however absent. It is is not possible to find, within a cell, an entity that collects information from various dynamical subsystems, processes such information and interferes into such subsystems, ensuring that they evolve in the required coordinated form. Instead, the coordination of various molecular processes comes as a result of autonomous interactions between them or, in other words, it is *self-organised.*

While biological cells and organisms are naturally available, they can also be seen as designed—not by an engineer, but by the process of biological evolution. Some basic underlying aspects of their operation could be already found in the original inanimate nature. However, through a long evolution history, extremely intricate self-organised systems of interacting chemical processes have emerged. A biological cell is a chemical micrometer-size reactor where thousands of different chemical reactions, sometimes involving only small numbers of molecules, proceed in a coordinated way. Such reactions are interconnected, when this is required by a function. Remarkably, however, they may be also non-interfering, even though confined to the same microvolume. On top of that, the entire system is accurately reproducing itself after every replication of the cell.

Definitely, this extreme level of molecular self-organisation comes at a high price. Because so many processes need to be packed into a tiny volume, some molecular components become shared by different mechanisms, so that they are optimised with respect to various functions. This presents a difficulty when biological systems are analysed. Another complication is that the actual living cells and organisms are the product of a unique and singular biological evolution, making it often difficult to say whether some property is essential for a specific function or it is accidental and results from a particular evolutionary path.

For those not involved in biology research, molecular biosystems may look very special—almost esoteric—and well outside their field. In a fact, they share the same basic physical principles with the systems of other origins and, in a more primitive form, the analogs of biological molecular processes can be identified in the inorganic nature too. Furthermore, artificial systems with similar functional properties can be designed.

Note that the development of new kinds of materials and physical systems already represents a major part of scientific research. Many physical processes that are broadly employed in modern technology can be found only in the rudimentary form, if at all, in the nature, so that the respective materials and devises need to be intensionally designed and produced. The semiconductors used in the electronics and computer chips are carefully fabricated and their performance is superficial as compared to those of naturally available materials where the phenomenon of semiconductivity is observed. Optical lasers are all artificially designed and manufactured, with no analogs found in the nature itself. In the chemistry field, most of the broadly used polymer materials are the products of intentional development and design.

One of the aims of our book is to emphasize that molecular processes with special functional properties, resembling to some extent biological organisation, need to be systematically developed and designed on a broad scale. In terms of future applications, introduction of such molecular processes into technology may well have an effect comparable to that of the invention of semiconductor circuits and laser devices. Noticing how much public attention and financial investment has been invested in the last decades to nanoscience and nanotechnology, it is surprising how little, in comparison, has been here done. Partly, this is explained by the fact that, in contrast to nanotechnology, the research on molecular self-organising systems has been historically fragmented and proceeding along several loosely connected lines.

Theoretical foundations have been laid, in terms of the thermodynamics of open systems, already by L. von Bertalanffy, E. Schrödinger and I. Prigogine. However, their contributions have only elucidated the governing principles of molecular self-organisation, without proposing specific design schemes. Taking optical lasers as a comparative example, one can notice that quantum statistical mechanics provides the basis for the design of such devices, but does not yet tell how to develop them. Subsequently, this theoretical approach has been explored using abstract reaction-diffusion models by A. Turing and I. Prigogine.

Experiments on self-organisation in molecular systems have been for a long time focused on the inorganic Belousov–Zhabotinsky reaction where persistent oscillations and various kinds of non-equilibrium wave patterns can readily be observed. This chemical system has however a model character. While it allowed one to demonstrate a rich spectrum of self-organisation phenomena, no practical applications could have been designed. More recently, oscillations and self-organised wave patterns could also be observed in catalytic reactions on solid surfaces. Nonetheless, such observations were mostly made in the context of understanding the mechanisms of heterogeneous catalysis.

Meanwhile, the attention had also become diverted from physical chemistry. Proceeding from his studies on quantum laser generation, H. Haken has noticed that spontaneous development of coherence and order can be a property of many different systems, including those of social origin. Hence, the term "synergetics" (from Greek: working together) has been coined for the new interdisciplinary field. In the US, a group of scientists in the Santa Fe institute was pushing forward the idea of artificial life; their theoretical studies were primarily focused on deeper understanding of evolution processes, but also contributed to the field of machine intelligence and networks science. At the same time, the concept of complex systems has emerged and became wide-spread in the context of interdisciplinary research.

While the suggestion of a molecular mechanism of heredity by E. Schrödinger has played a decisive role in the subsequent discovery of the genetic code by J. Watson and F. Crick and thus had a profound effect, his ideas on physical mechanisms of self-organisation remained for a long time discussed by physicists, rather than by biologists themselves. Despite its impressive experimental achievements, molecular biology continued to rely, in its theoretical interpretations, on the concepts of classical chemical kinetics. It was not, until recently, much affected by the developments in the theory of complex systems that became gradually dominating the interdisciplinary research. The situation in molecular and cell biology has however changed within the last decade. High-resolution microscopy methods for in vivo monitoring of chemical processes in biological cells have become available. Thus, various self-organisation processes in living cells could be observed and this has stimulated the respective modeling and theoretical research.

Perhaps, the time has come to summarise, in a historical perspective and from the viewpoint of physical chemistry, what has been done in the studies of self-organisation processes in molecular systems. It needs also to be discussed in what directions the future studies can proceed and what applications become feasible. The present book does not however intend to systematically cover all relevant aspects— our attention is only paid to selected topics whose choice reflects largely the personal interests of the authors. Nonetheless we hope that this work can contribute to the broad survey of the history and the current status of the research on self-organization in molecular systems.

# Chapter 2
# Thermodynamics of Open Systems

In his speech at the Prussian Academy of Sciences in 1882, the scientist Emil du Bois-Reymond, well-known at that time, has concluded: "Chemistry is not a science in the sense of the mathematical description of Nature. Chemistry will be the science in this highest human sense only if we would understand the forces, velocities, stable and unstable equilibria of particles in a similar manner as the motion of stars" [1]. The thermodynamic theory for "stable equilibria" in closed systems was essentially developed during the second half of the 19th century and completed in 1905 by W. Nernst through the formulation of the Third Law of thermodynamics, while the basis for "forces and velocities" had to wait for the advent of quantum mechanics during the first decades of the 20th century. The description of "unstable equilibria" (which among other underlies all phenomena of structure formation in biological systems) became however only accessible in the middle of the 20th century in the framework of thermodynamics of open systems.

According to the Second Law of thermodynamics, all closed physical systems tend to reach the state of thermal equilibrium characterized by the minimum of free energy. But this is obviously not the case in biology. A living biological organism is not in the state of thermal equilibrium which can only be reached when the organism is dead. Quite in contrast, the degree of organisation, i.e. the order, of a bio-organism increases over time in the process of its development from the initial cell.

At the beginning of the 20th century, many have suspected that peculiar "vitalistic" forces, valid only in biology, exist. If such forces were indeed found, this would have however made thermodynamics and statistical physics not universally applicable, thus undermining the unique physical picture of the world. Therefore, efforts have been started to reconcile biology to the physical laws. Ludwig von Bertalannfy, an Austrian philosopher, biologist, system scientist and psychologist, was the first to address this paradox. In 1926, von Bertalannfy had finished his study of philosophy and art history with a doctoral degree at the University of Vienna and, 2 years later, published his first book on theoretical biology, *Kritische Theorie der Formbildung* (Critical Theory of Form Development), soon followed by

© Springer International Publishing AG 2017
A.S. Mikhailov and G. Ertl, *Chemical Complexity*, The Frontiers Collection,
DOI 10.1007/978-3-319-57377-9_2

two volumes of *Theoretische Biologie*. In his article [2] with the title "Der Organismus als Physikalisches System Betrachtet" (The Organism Considered as a Physical System), von Bertalannfy offered in 1940 deep insights in the physical nature of biological phenomena.

As he noted, a biological organism shows properties similar to those of equilibrium systems. Indeed, the composition of a cell or of a multi-cellular organism is maintained over time and recovered after perturbations. But, although equilibrium of elementary subsystems can be found within it, the organism as a whole cannot be considered as being in the state of equilibrium. This is because we deal here not with a closed, but with an open system. A system should be called "closed" if there is no material entering it from outside and leaving it. In an open system, on the other hand, supply and release of materials take place.Thus, he wrote:

> The organism is not a static, isolated from the exterior, system that always contains identical components. Rather, it is an open system in a (quasi) stationary, or steady, state that retains its mass relations under permanent exchange of substances and energies building it, the state where some components persistently arrive from outside while other components are persistently leaving.

At that time, physical chemistry was essentially limited to the analysis of reaction processes in closed systems. As von Bertalannfy remarked, open systems are not of much theoretical interest in the field of pure physics. One can however easily imagine that, for example, in the reaction $a \rightleftharpoons b$ the product $b$ of the reaction running from the left to the right is permanently removed, whereas the substrate $a$ is permanently supplied. This is exactly so with a chemical flow reactor working under steady-state conditions, and such systems also play a fundamental role in biology.

L. von Bertalannfy stressed that, while there are similarities between stationary states ("unstable equilibria") in open systems and the equilibrium in closed systems, the physical situation is principally different in these two cases:

> True chemical equilibria in closed systems rely on reversible reactions [...]; they are furthermore a consequence of the Second Law [of thermodynamics] and are defined through the minimum of free energy. In open systems, the steady state as a whole and, eventually, also many elementary reactions are not however reversible. Moreover, the Second law is applicable only to closed systems, it does not determine the steady states. A closed system *must*, on the basis of the Second Law, finally go into a time-independent state of equilibrium [...] where the relationship between phases stays constant. An open chemical system *can* (when certain conditions are satisfied) finally go to a time-independent steady state. The characteristic property of this state is that the system, as a whole and also in view of its (macroscopic) phases, keeps itself constant through an exchange of elements (the so-called dynamic equilibrium).

Thus, according to von Bertalannfy, the paradox is only superficial. All laws of physics are in principle applicable to biological systems, provided that their necessary conditions are satisfied. The Second Law of thermodynamics is not violated in biology—it does not hold for biological organisms because they represent open systems and thus the applicability conditions are not satisfied. The steady state of an open system can change when flows passing through it are modified.

There is one further important consequence of this analysis. To maintain a closed system at equilibrium, no work is needed, but work cannot be also performed by a system in such a state. As von Bertalannfy remarks, a dammed mountain lake contains much potential energy, but, in absence of the outgoing flow, it cannot power a motor or a turbine. To generate work, the system needs to be under transition to an equilibrium state. To keep the system over a long time under a transition, one has to engineer it like a water power station, supplying new material whose energy is used to produce work. Furthermore,

> Persistent work generation is therefore not possible in a closed system that rapidly transits to the equilibrium state, but only in an open system. The apparent "equilibrium", in which an organism finds itself, is therefore not a true - and hence not capable of work - equilibrium state, but rather a dynamic pseudo-equilibrium. It is hold at a certain separation from the true one and is therefore able to produce work. On the other hand, it also permanently needs new energy supply to ensure that a distance from the true equilibrium state remains maintained.

After the second world war, von Bertalannfy has moved to the US where his interests became shifted from theoretical biology to other fields. He has founded the general systems theory, analysing organization principles in systems of various origins, and was also involved in cancer studies and psychology research. Although his analysis of physical principles in biological organisms was brilliant, it was mostly confined to the conceptual level. Being not a mathematician or a theoretical physicist, von Bertalannfy could not further cast his ideas into an adequate mathematical form.

At about the same time, biology problems attracted attention of Erwin Schrödinger, famous for his discovery of quantum mechanics together with Werner Heisenberg. In emigration in Ireland, he had given in 1943 several lectures in the Dublin Institute for Advanced Studies which were published one year later as a book with the title *What is Life? The Physical Aspect of a Living Cell* [3]. The lectures were attended by an audience of about four hundred, both physicists and biologists. Despite his brilliant command of mathematics, Schrödinger almost did not use it during these lectures.

The lectures were focused on unveiling possible mechanisms of genetic inheritance in biological organisms. The puzzle was to explain how genetic information is reliably transferred in large amounts from one generation to another, despite the apparent frailty of biological organisms. His conclusion was that the information should be stored at the molecular level, in large molecules that may resemble an "aperiodic crystal". Somewhat later, this conjecture became indeed confirmed when the genetic DNA code was discovered by J. Watson and F. Crick. However, one of the lectures was devoted to thermodynamic aspects of the living cell.

Schrödinger begins by noting that biological organisms are open systems and therefore the Second Law of thermodynamics is not applicable to them. It is not clear whether he was familiar at that time with the work by von Bertalannfy; the book contains no reference to it. During the war, Schrödinger was in relative isolation in Ireland and did not also know about some contemporary experimental developments in biophysics. Similar to von Bertalannfy, he asks a question whether the laws of physics hold in biology. His answer is that "the living matter, while not eluding the "laws of physics" as established up to date, is likely to involve "other laws of physics" hitherto unknown, which, however, once they have been revealed, will form just integral a part of this science as the former."

How does a biological organism avoid decay? — By eating, drinking and breathing (or assimilating in the case of plants). Indeed, one of the fundamental concepts in biology is that of metabolism. This word, translated from Greek, means change or exchange. In German literature, the term *Stoffwechsel*, i.e. exchange of material, is used as a synonym of metabolism. Obviously, the metabolism is essential for keeping the cell alive, but this does not simply mean the exchange of molecules.

The characteristic feature of a biological organism is a high degree of order within it. This order is maintained despite large variations in the environment. It can also increase with time, as, for example, in the process of development of a multi-cellular organism from a single initial cell. In thermodynamics, entropy serves as the measure of disorder and this means that, when order of an organism is increased, its entropy content must become lower.

Assume that $S$ is the entropy contained within a system. Then, its rate of change is given by the sum of two terms

$$\frac{dS}{dt} = \sigma + j^s. \tag{2.1}$$

The first of them, $\sigma$, is the rate of entropy generation within the system. The Second Law of thermodynamics implies that this term is always non-negative and vanishes in the state of thermal equilibrium. The second term, $j^s$, is the rate of exchange of entropy between the system and its environment. It is given by the difference $j^s = j^s_{in} - j^s_{out}$ of incoming and outgoing entropy fluxes. Note that this second term should also include the exchange of entropy between the system and the thermal bath.

If the amount of entropy arriving per unit time to an open system is smaller than the amount exported by it per unit time, the total entropy flux $j^s$ in Eq. (2.1) becomes negative. Furthermore, if this negative flux is sufficiently strong, it can prevail over the entropy production rate, so that we would have $\sigma + j^s < 0$. But this means that, under such conditions, $dS/dt < 0$ and thus the entropy content of the system *decreases* with time.

Moreover, if the two terms exactly balance one another, $j^s = -\sigma$, the entropy content of the system remains constant with time. However, this is obviously not the state of thermal equilibrium, since entropy continues to be produced. Instead, the open system will find itself in this case in the steady state of flow equilibrium that we have already discussed above.

In his book, Schrödinger suggests that, instead of the entropy, its reverse $\Psi = -S$ should be rather used in such arguments. The "negative entropy" $\Psi$ characterizes the degree of order in the system and, for it, an analog of Eq. (2.1) holds,

$$\frac{d\Psi}{dt} = -\sigma + j^\psi_{in} - j^\psi_{out}. \tag{2.2}$$

The order of an organism would increase if the amount $j_{in}^{\psi}$ of negative entropy consumed by it per unit time exceeds the amount $j_{out}^{\psi}$ of negative entropy exported within the same time. To maintain an organism in the non-equilibrium steady state, the flow of incoming negative entropy has at least to overcome entropy production within it,

$$j_{in}^{\psi} = \sigma + j_{out}^{\psi} \geq \sigma. \tag{2.3}$$

Thus, to keep itself alive, a biological organism needs to "feed on negative entropy" or, in other words, it should continuously import "order" from an external source. The rate of such import should be high enough, so that the internal entropy production is counter-balanced. The conclusion by Schrödinger was that, in biological organisms, order is not created—it is only imported from outside. Indeed, "Order from Order" was the title of the section where such questions were discussed.

However, the concept of "negative entropy" has not become popular, partly because this is just the common entropy taken with the opposite sign. In the *Notion* to the respective chapter of the book, Schrödinger admitted that his remarks on negative entropy were met with doubt and opposition from his physicist colleagues. He said that he had chosen to talk about negative entropy in his lecture only to explain the situation in a simpler way for a broad audience. Instead, he should have talked about the free energy of a biological organism.

Indeed, thermodynamic free energy is defined as $F = E - TS$ where $E$ is the energy and $T$ is the temperature of a system. Assuming that (internal) energy $E$ is not consumed or released inside a system, the balance equation for free energy can be written as

$$\frac{dF}{dt} = -T\sigma + j_{in}^{F} - j_{out}^{F}. \tag{2.4}$$

Hence, in the steady state of an open system, the influx of free energy should be sufficiently high,

$$j_{in}^{F} \geq T\sigma. \tag{2.5}$$

It has to overcome the effect of persistent entropy production in the steady state.

Hence, there is indeed something essential that should be received by all of us with food. This is not however the simple energy, but rather the thermodynamic free energy contained in it. A minimum amount of free energy has to be consumed every day in order that a biological organism survives.

After the war, Schrödinger returned to Austria where his research was again focused on physics. The book *What is Life?* was his only major publication where problems of biology were discussed.

The third main contribution to studies in thermodynamics of open systems came from Ilya Prigogine. He was born in Russia and emigrated as a child with his parents first to Germany and then to Belgium. In 1941 he has received a doctoral degree from Université Libre de Bruxelles and, in 1947, published his first book *Ètude Thermodynamique des Phènoménes Irreversibles* [4]. In contrast to von Bertalanffy, a theoretical biologist, or to the physicist Schrödinger, he was a physical chemist. Apparently without being aware of previous work, Prigogine has not only reinvented the principles of thermodynamics of open systems, but also systematically built up this theory as applied to specific chemical systems.

Reviewing in 1949 the book [4] by Prigogine for the *Nature* magazine [5], von Bertalannfy wrote: "Since 1932, the present reviewer has advanced the conception of an organism as an "open system". So far, physical chemistry was concerned almost exclusively with reactions and equilibria in closed systems, while living organisms are open systems, maintaining themselves in a continuous exchange of materials with environment. [...] Prigogine's work, devoted to the extension and generalization of the thermodynamic theory, is of outstanding importance for physics, as well as for biology. As the author states, "classical thermodynamics is an admirable, but *fragmentary* doctrine; this fragmentary character results from the fact that it is applicable only to states of equilibria in closed systems. *Therefore, it is necessary to establish a broader theory, comprising states of non-equilibrium as well as those of equilibrium.*" Starting, on one hand from, from the concept of open systems and, on the other, from thermodynamics of irreversible processes [...], Prigogine derives the generalized thermodynamics for the whole realm of physical chemistry, including chemical reactions, electrochemistry, polythermic systems, diffusion, thermoelectricity, etc. [...] The new thermodynamics shows that it is necessary not only for biological theory to be based upon physics, but also that biological points of view can open new pathways in physical theory as well."

Over his long life in science, Prigogine has worked on a broad spectrum of topics, including not only thermodynamics of open systems, but also various aspects of non-equilibrium pattern formation, foundations of physical kinetics and quantum statistics, and interdisciplinary applications to social and technological problems. He was a prolific writer and has left a series of excellent books. In 1977, Prigogine was awarded with the Nobel prize in chemistry "for his contributions to non-equilibrium thermodynamics, particularly the theory of dissipative structures".

Below we show, following [6], how thermodynamics of open systems can be applied to specific examples.

Suppose that a system is connected to just one other system that is large and plays a role of the thermal bath (Fig. 2.1). Such a system is closed in the thermodynamic sense and can only undergo a transition to the equilibrium state. Note that, during

**Fig. 2.1** A system in contact with the thermal bath

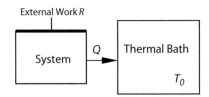

**Fig. 2.2** A system in contact with two thermal baths at different temperatures

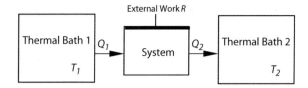

the transient, the system can still exchange heat with the thermal bath, so that the internally produced entropy is exported and the entropy content of the system is decreased. This indeed takes place if, for example, the initial temperature of a closed system is higher than that of the thermal bath, so that it gets cooled.

The situation is however different if a system is connected to two (or more) thermal baths with different temperatures, as shown in Fig. 2.2. Now, the system cannot settle down to any definite equilibrium state. Even if its state is stationary (steady), heat would continue to arrive from the bath with the higher temperature $T_1$ and become released into the colder bath at the temperature $T_2$. To estimate the amounts of entropy transported with the heat flows, a simple argument can be used. If the heat $dQ_1$ flows into the system from the thermal bath at temperature $T_1$, the same amount is leaving that bath. Thus, the bath entropy becomes decreased by $dS_1 = dQ_1/T_1$ and the same amount of entropy arrives with the heat flow into the considered middle system. On the other hand, if an amount of heat $dQ_2$ enters the second thermal bath which is at temperature $T_2$, it brings with it the entropy $dS_2 = dQ_2/T_2$ and, therefore, the same amount of entropy is leaving the system connected to that bath.

Thus, the Second Law of thermodynamics takes here the form

$$\frac{dS}{dt} = \sigma + \frac{1}{T_1}\frac{dQ_1}{dt} - \frac{1}{T_2}\frac{dQ_2}{dt} \qquad (2.6)$$

Moreover, the First Law of thermodynamics implies that the balance equation for the internal energy $E$ of the system should hold,

$$\frac{dE}{dt} = \frac{dQ_1}{dt} - \frac{dQ_2}{dt} + \frac{dR}{dt}. \qquad (2.7)$$

where $R$ is the work produced by the system. If the system is changing its volume $V$, the rate of work generation (or the generated power) is given by $dR/dt = -p(dV/dt)$ where $p$ is pressure.

Because the system is away from equilibrium, it does not generally need to approach a steady state and can, as well, perform periodic oscillations or even show more complex dynamics. Suppose however that such a stationary state exists. Then, $dE/dt = dS/dt = 0$ and, from Eq. (2.6) we find

$$\frac{dQ_2}{dt} = T_2\sigma + \frac{T_2}{T_1}\frac{dQ_1}{dt} \qquad (2.8)$$

This expression can be further substituted into Eq. (2.7) and the rate of work generation can be determined,

$$\frac{dR}{dt} = -T_2\sigma + (\frac{T_2}{T_1} - 1)\frac{dQ_1}{dt}. \tag{2.9}$$

As we see, this open system can operate as a *heat engine* and perform work. Because the entropy production rate $\sigma$ is positive, the maximum efficiency is reached when $\sigma \to 0$, i.e. if the system approaches equilibrium. Then, the efficiency coefficient is

$$\eta = \frac{dR/dt}{dQ_1/dt} = \frac{T_2}{T_1} - 1. \tag{2.10}$$

Remarkably, this is indeed the maximum energetic efficiency of the Carnot cycle.

In the above arguments, we have assumed that only heat could be exchanged, but the material composition of the system remained fixed. In chemistry and biology, a situation would however often be found where material flows are taking place, whereas temperatures are not different for different parts. Typically, a chemical reaction is taking place inside a system (the *reactor*) where molecules $X$ are reversibly converted into some other molecules $Y$. Again, if the system is closed, an equilibrium state with certain stationary concentration of species $X$ and $Y$ is eventually reached. Now we assume that the considered chemical reactor is connected to two large external systems (or *chemostats*) where molecules $X$ or $Y$ are contained and their concentrations are maintained constant (Fig. 2.3). Moreover, chemical potentials $\mu_X$ and $\mu_Y$ in the chemostats are different and $\mu_X > \mu_Y$. The chemostats and the system are kept at a constant pressure, so that the Gibbs free energy should be employed. The system is also connected to a thermal bath at temperature $T_0$.

The First Law of thermodynamics then reads

$$\frac{dE}{dt} = \mu_X j_X - \mu_Y j_Y + \frac{dR}{dt} - \frac{dQ}{dt} \tag{2.11}$$

**Fig. 2.3** Open system connected to two chemostats

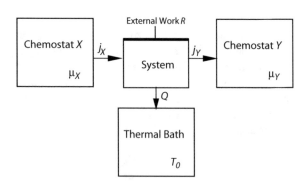

where $j_X$ is the number of molecules $X$ arriving per unit time into the system and $j_Y$ the number of molecules $Y$ leaving it per unit time, and the last term takes into account the heat flow into the thermal bath.

The Second Law implies that

$$\frac{dS}{dt} = \sigma + s_X j_X - s_Y j_Y - \frac{1}{T_0} \frac{dQ}{dt} \qquad (2.12)$$

where $s_X$ and $s_Y$ are amounts of entropy per molecule $X$ or $Y$ in the respective chemostats; they are given by $s_X = -\partial \mu_X / \partial T$ and $s_Y = -\partial \mu_Y / \partial T$.

Additionally, mass balance equation needs to be written. Assuming that, in the considered chemical reaction, one molecule $X$ is converted into one molecule $Y$, that is we have $X \rightleftharpoons Y$, the mass balance implies that the number of molecules entering the reactor per unit time is equal to the number of molecules leaving it, i.e. $j_X = j_Y = j$.

Such a flow reactor will be maintained away from thermal equilibrium as long as there is a flow of molecules passing through it. Generally, oscillations and various wave patterns can develop inside it. It may happen however also that a stationary (steady) state becomes formed, so that $dE/dt = dS/dt = 0$. In contrast to thermal equilibrium state, entropy continues to be produced under stationary non-equilibrium conditions. Additionally, some entropy arrives together with substrate molecules $X$. All such entropy has to be exported away from the reactor. According to Eq. (2.12), it is exported together with the heat flow into the thermal bath and together with the product molecules into the respective chemostat.

From Eq. (2.12), we find that, in the steady state, the heat flow into the thermal bath is

$$\frac{dQ}{dt} = T_0 \sigma + T_0 (s_X - s_Y) j. \qquad (2.13)$$

Substituting this, in the steady state, into Eq. (2.11), mechanical work performed per unit time by such chemical engine can be determined,

$$\frac{dR}{dt} = -T_0 \sigma + [T_0 (s_X - s_Y) - (\mu_X - \mu_Y)] j. \qquad (2.14)$$

As in the case of the heat engine, the work is maximal when approaching equilibrium, i.e. for $\sigma \to 0$. The maximum possible efficiency of the chemical engine is

$$\eta = \frac{dR/dt}{(\mu_X - \mu_Y) j} = \frac{T_0}{\mu_X - \mu_Y} \frac{\partial}{\partial T} (\mu_X - \mu_Y) - 1. \qquad (2.15)$$

The possibility to operate as an engine, persistently producing mechanical work, is a remarkable property of open flow systems, both thermal and chemical. There is however also another important property of such systems revealed by this analysis. Their entropy content is controlled by a balance between the entropy production inside the system and the entropy flows through the system's boundaries. By adjusting such flows (which can be done by the system itself) the entropy of the steady state can be maintained at any desired level and thus the states with different degrees of order can be reached and kept. Thus, an open system is to a certain extent autonomous, or *self-organized*. Such autonomy becomes even more evident in spontaneous development of oscillations, dissipative structures and traveling waves in open systems far from thermal equilibrium.

Our examples also illustrate, in agreement with the conclusions by E. Schrödinger, that the order developing within an open system is a result and a manifestation of the order already present in the world around it. Indeed, a thermal engine can only operate if, in the environment, there are two thermal baths with different temperatures which are in contact with it. In a similar manner, the pre-requisite for the operation of a chemical engine is that two molecular reservoirs with different chemical potentials exist. Of course, this means that the world into which such an open system (or a biological organism) is placed is far from thermal equilibrium itself. Open systems do not create the order, but rather receive it from the outside world.

The question is then what is the source of order in the surrounding world. If we look at the Earth as a whole, we can notice that it is a kind of a "heat engine" itself. With the light, our planet receives heat from the sun that can be viewed as a thermal bath with at a high temperature. On the other hand, heat is released by the Earth as radiation into the open space filled with the background cosmic radiation at the low temperature of only a few K. This flow leads to the atmospheric and other activity on the Earth and is the ultimate source of the order on which all biological organisms "feed". When nuclear fuel in the sun will become exhausted and the sun will stop to shine, life will disappear too.

Before we conclude the present chapter, some comments concerning relaxation to equilibrium in closed systems should be made. During the transient, while thermal equilibrium is being approached but has not yet been reached, closed systems may exhibit similarities to open systems and, for instance, can perform work. In most cases, such transients are rapid and do not deserve special investigation. There are however situations when these transients become exceptionally long. A system may include a depot where energy or material are initially stored and slowly released and made available for other parts of it. From the viewpoint of the rest of the system, such depot would obviously represent an external source and, therefore, the analysis for open systems will also hold in this case.

# References

1. E. du Bois-Reymond, cited after: W. Ostwald, Z. Phys. Chem. 1, 1 (1887)
2. L. von Bertalannfy, Naturwissenschaften **33**, 34 (1940)
3. E. Schrödinger, *What is Life? A Physical Aspect of a Living Cell* (Cambridge University Press, Cambridge, 1944)
4. I. Prigogine, *Étude Thermodynamique des Phénomènes Irréversibles* (Paris: Libr. Dunod; Liège: Éditions Desoer,1947)

5. L. von Bertalannfy, Nature **163**, 384 (1949)
6. P. Glansdorff, I. Prigogine, *Thermodynamic Theory of Structure, Stability and Fluctuations* (Wiley, New York, 1971)

# Chapter 3
# Self-assembly Phenomena

*And the Earth was without form, and void.*
Old Testament, Genesis 1:2

According to modern cosmology, the Universe was born as a burst of energy. The initial temperature of the drop was so high that even atoms could not exist—they have appeared at a later stage, and then various molecules became established too. These constituents have gradually assembled to form stars and planets and, at some point, life on the Earth has emerged. The initial great disorder in the Universe is manifested in the distribution of the cosmic microwave background radiation that could be measured by the COBE satellite. Random localization of recorded temperature fluctuations reveals the statistical nature of the structure at the early stage.

As the Universe was gradually acquiring its shape, not only its structural units—elementary particles, atoms and molecules—have emerged, but the laws that govern interactions between them have arisen too. Remarkably, in his inaugural lecture [1] at the University of Zürich in 1921 E. Schrödinger has already pointed out that "The physical research has unequivocally demonstrated that, at least for the overwhelming majority of observed phenomena whose regularity and continuity led to the formulation of the postulate of general causality, the common root for the strict laws has to be sought in the accident." Such dramatic evolution has been caused by persistent expansion of the Universe. Since it is a closed system, expansion has lead to temperature decrease, extremely rapid when the Universe was just been born. Through abrupt cooling, structural order has developed within it.

While the history of the Universe provides the most impressive example of self-assembly, similar effects can be found, on different length and time scales, in various physical systems. Usually, such systems are not isolated, but interact with a thermal bath. In the course of time, they tend to approach the state of equilibrium at the temperature of the bath. If initially such a system is in a hot state, heat and entropy will flow from it into the bath, so that the order of the system can become enhanced.

© Springer International Publishing AG 2017

A.S. Mikhailov and G. Ertl, *Chemical Complexity*, The Frontiers Collection,
DOI 10.1007/978-3-319-57377-9_3

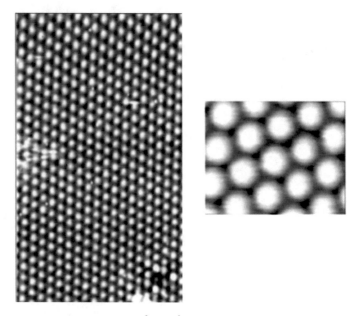

**Fig. 3.1** (*Left*) Al(111) surface area (46 Å × 71 Å) imaged with atomic resolution by ATM. (*Right*) Enlargement for the area with the size of 13 Å×9 Å. H. Brune and G. Ertl (unpublished)

Alternatively, the temperature of the bath can be decreased, so that the system becomes cooled. In this manner, perfectly ordered structures can be produced even on the atomic scales.

Figure 3.1 shows, as an example, a section from the (111) surface of an aluminium single crystal recorded by scanning tunneling microscopy (STM) with atomic resolution. This perfectly ordered structure had been grown by cooling down the material from the melt, i.e. a highly disordered state.

According to the Second Law of thermodynamics, a system interacting with the thermal bath evolves in such a way that its free energy is gradually decreased, reaching its minimum at the equilibrium state. The free energy $F$ of a system is given by the equation $F = E - TS$ where $E$ is the internal energy and $S$ is the entropy of the system. At high temperatures $T$ the term $-TS$ will prevail leading to disorder, while at lower temperatures the equilibrium will be dominated by the minimum of $E$ that may be associated with an ordered configuration, such as in a crystal.

The free energy $F$ is a function of the internal variables, such as the coordinates of individual elements forming the system. By definition, this function should have a minimum in the equilibrium state. However, the multi-dimensional landscape of the free energy will often not only exhibit the global minimum corresponding to the true thermal equilibrium, but also contain a number of local minima reflecting the existence of metastable states. Such additional minima are separated by energy barriers characterizing kinetic constraints. Moreover, two minima can also have the same depth so that two physical phases coexist.

**Fig. 3.2** Lattice gas of
oxygen adatoms on
ruthenium single crystal
surface. Image size is 80 Å
× 80 Å. Reproduced from [2]

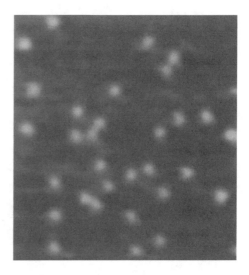

When temperature is decreased, phase transitions take place. A system of interacting molecules behaves as gas at high temperatures, but, as temperature is decreased,
it undergoes condensation first into the fluid and then into the crystalline state. While
condensation phase transitions are classical and they have been studied for a long
time, more recently it has become possible to observe and investigate them with
atomic resolution and in real time.

Figure 3.2 shows a snapshot taken at the temperature of 300 K with STM from
a ruthenium single crystal surface onto which a small amount of oxygen (O) atoms
has been deposited [2]. The free valencies of the topmost metal atoms gave rise to
strong bonds with the oxygen atoms which nonetheless could occasionally hop to
neighbouring sites of the periodic lattice formed by the metal substrate.

By employing fast STM, not only positions of individual oxygen adatoms could be
resolved in the experiments [2], but also hopping of O atoms between the sites could
be directly observed and their mean residence times could be determined. While
randomly moving around, oxygen atoms had the mean residence time at a specific
site of 0.06 s if there were no other oxygen adatoms around them. The residence
time rose however to about 0.2 s if there was a neighbour oxygen atom separated by
the distance of two lattice lengths of the substrate, indicating the presence of weak
attractive interactions between oxygen adatoms.

Because of attractive interactions, formation of two oxygen phases, quasi-gaseous
and quasi-crystalline, took place as surface concentration of oxygen was increased.
Figure 3.3a, b shows two snapshots of the ruthenium surface separated by time interval of 0.17 s. Here, the mean surface coverage by oxygen, i.e. the fraction of lattice
sites occupied by O atoms, is $\theta = 0.09$. Additionally, Figs. 3.3c, d show the same
snapshots in a schematic representation where only the positions of O adatoms are
displayed. The atoms that have changed their positions within the considered time
interval are shown by light gray circles. One can see that surface oxygen has formed

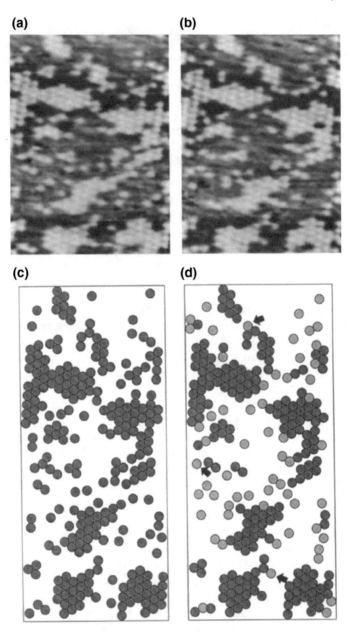

**Fig. 3.3** Coexistence of gaseous and crystalline phases of oxygen on single crystal ruthenium surface. The same parameters as in Fig. 3.2, but surface coverage of oxygen is increased to $\theta = 0.09$. The image size is 80 Å × 190 Å. The two snapshots **a** and **b** are separated by the time interval of 0.17 s. In panels **c** and **d**, only the positions of O adatoms are displayed. *Light grey circles* in panel **d** show O adatoms that have changed their position within 0.17 s. Reproduced from [2]

dense islands inside which the maximum possible coverage is reached. Such crystalline islands have irregular shapes and different sizes. The areas between them are occupied by the gas of single O adatoms. By comparing Fig. 3.3a, c and b, d, it can be observed that, in such areas, oxygen adatoms are mobile. Moreover, it can be also seen that the islands can change their configurations by attaching new adatoms from the gas phase and also by detachment of O atoms from them. On the average, the two adsorbate phases seem to be at equilibrium one with another, similar to solid ice and water vapor in three dimensions.

Such a surface system can also be viewed as a mixture of two kinds of particles, A and B, with particles A corresponding to oxygen adatoms and particles B corresponding to vacancies, i.e. the empty lattice sites. The concentrations of such particles can be characterized by their surface coverages $\theta_A$ and $\theta_B$; in the considered case we have $\theta_A = \theta$ and $\theta_B = 1 - \theta$. The free energy $F = E - TS$ of the mixture consists of the internal energy and the entropic contribution, it should have its minimum in the equilibrium state. Because of attractive short-range interactions between particles A, the lowest total internal energy $E$ of the system will be found if all particles A aggregate in a single large island inside which all available lattice sites are occupied and particles B occupy the rest of the system. On the other hand, the entropy $S$ of the mixture will be maximal if particles A and B are randomly and statistically uniformly distributed over the entire surface.

If temperature $T$ is high, the free energy is dominated by the entropy term and the equilibrium is reached at the uniform distribution of the particles. If temperature is low, the internal energy prevails and, at equilibrium, the system is found in the phase-separated state. Note that, unless $T = 0$, some of the particles A should be also found outside of dense islands. Indeed, there is then a certain probability of thermal "evaporation" events when one of the particles A at the boundary of the island detaches from it (such events can be indeed seen in Fig. 3.3). Moreover, some particles B may be also found, at finite temperatures, inside the regions occupied by particles A.

The behavior of such binary mixture can be rationalized in its phase diagram schematically drawn in Fig. 3.4. In this diagram, temperature is plotted along the vertical axis. The variable plotted along the horizontal axis is the *average* coverage $\bar{\theta}$, i.e. the total number of available particles A divided by the total number of surface lattice sites. Phase separation occurs within the central region bounded by the bold line in Fig. 3.4. This region is always entered as the system is further cooled. Inside

**Fig. 3.4** Schematic phase diagram for a binary mixture

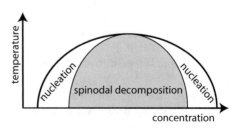

it, the system is divided into two distinct phases I and II, occupied predominantly by particles A or B at certain local coverages $\theta_I$ and $\theta_{II}$. Because the total number of particles A is fixed, we always have $\bar{\theta} = \sigma\theta_I + (1 - \sigma)\theta_{II}$ where $\sigma$ is the fraction of surface sites occupied by phase I.

Using the phase diagram, one can also analyze what happens if, by keeping the temperature constant, we begin to increase the total number of particles A. At small coverages $\bar{\theta}$ we are on the left side from the bold line in Fig. 3.4 and thus in the region of the single gas-like phase, like in Fig. 3.2. As soon as this line gets crossed, phase separation into a diluted (gaseous) and condensed phases, such as in Fig. 3.3, takes place. Upon further addition of particles, the amount of the condensed phase is growing. Eventually, the phase separation boundary is crossed on the right side and the system becomes statistically uniform again, occupied completely by the condensed phase. Note that separation into gas-like and condensed phases can only be found if temperature is small enough. Above the critical point, corresponding to the top of the phase separation curve, there is a gradual transition from the uniform gas-like to the uniform condensed phases as the number of particles is increased.

According to the theory, the system should separate into only two regions with different phases in the final equilibrium state. Indeed, if several such regions are present, the free energy is larger because the total length of the interface boundary is longer and more particles A have as their neighbors particles B. Therefore, such regions should gradually merge until the surface is divided into two parts. In the experiments with oxygen adsorbed on the ruthenium surface [2], the growth of islands has not however been observed. While the islands were changing their configurations, the tendency of their sizes to increase could not be seen. A possible explanation might be that, when an island of oxygen adatoms is formed, this induces strain in the substrate and the elastic deformation energy increases with the island size. This effect can limit the growth of the islands and stabilize their size (see also the discussion of micro-phase separation later in this chapter).

Suppose that the system has been prepared in the uniform state and then it evolves from such initial state to its thermal equilibrium at a given temperature. Such evolution represents a *kinetic* process and it will depend on how the system kinetics is organized (in contrast to the final equilibrium state that cannot depend on any kinetic constants). Some general conclusions about the kinetics of phase transitions can nonetheless be made.

Next to the phase separation boundary, the transition through a nucleation process takes place (Fig. 3.4). In order to initiate phase separation from the homogeneous phase, a sufficiently strong local perturbation—the *critical nucleus*—should be first created. On the left side, such perturbation is a small condensed island inside the dilute phase, whereas it represents a small void inside the dense phase on the right side. The characteristic size of the critical nucleus increases as the phase separation boundary is approached. If a perturbation exceeds in its size the critical nucleus, it grows; if it is smaller, it shrinks. The growing nuclei compete and, through the process known as Ostwald ripening, final transition to macroscopic phase domains takes place. Sufficiently strong perturbations, triggering the phase transition process, can arise from thermal fluctuations. They can be also created by local heterogeneities, such as

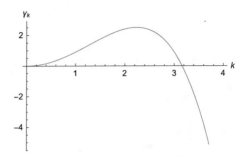

**Fig. 3.5** Rates of growth of perturbations with different wavenumbers under spinodal decomposition

structural defects or micro-inclusions. The control of nucleation process provides the basis for the formation of nanoparticles with well-defined size distributions, leading to the wide applications of nanotechnology.

Within the dark central region in Fig. 3.4, there will be no longer a barrier for nucleation and therefore spontaneous phase separation known as *spinodal decomposition* [3] occurs there. Suppose that at time $t = 0$ a small perturbation in the form of a plane wave with a certain wavenumber $k$ and amplitude $A_k(0)$ has been added to the uniform state. The initial evolution of such weak perturbation should be governed by a linear equation and its solution will have the form of a plane wave with the time-dependent amplitude

$$A_k(t) = A_k(0) \exp(\gamma_k t) \tag{3.1}$$

where $\gamma_k$ is the rate of growth of the perturbation mode with the wavenumber $k$.

Under spinodal decomposition conditions, $\gamma_k$ usually depends on $k$ as schematically shown in Fig. 3.5. All perturbation modes with the wavenumbers in the interval $0 < k < k_{max}$ are growing and contribute to the instability of the uniform state; the maximum rate of growth is found for a perturbation mode with a certain wavenumber $k_c$. Any initial perturbation can be decomposed into a sum of plane waves with different wavenumbers $k$. In the linear regime, amplitudes of pertubation modes will grow independently, each at its own rate. Because the growth is exponential, the evolving structure soon becomes dominated by the modes with wavenumbers $k \simeq k_c$ and therefore has the characteristic length scale $l = 2\pi/k_c$. In the middle of the phase disgram, this structure has the form of an interconnected labyrinthine pattern. Later on, nonlinear effects leading to interactions between the modes will come into play. According to the theory of spinodal decomposition, the morphology of the developing pattern does not qualitatively change then, but its characteristic scale $l(t)$ increases with time. At the end, two macroscopic domains with different phases become established.

The process of spinodal decomposition is illustrated in Fig. 3.6 where the results of a computer simulation [4] for a surface system are shown. In this simulation, a random initial distribution of particles at coverage $\theta = 0.5$ and with weak interatomic attractive interactions between the particles of strength 4 K was allowed to start to

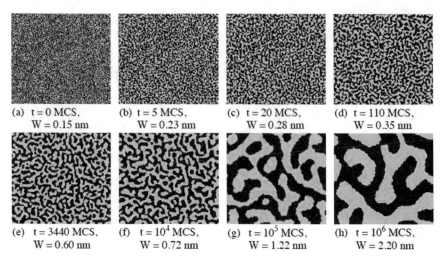

**Fig. 3.6** Computer simulation of spinodal decomposition in a surface system. Snapshots at different time moments are shown; $W$ is the characteristic length scale of a pattern. Reproduced from [4]

diffuse at $t = 0$. A labyrinthine pattern with the characteristic nanometer scale range had emerged and it was continuously growing while increasing its characteristic wavelength.

To observe this different kind of phase separation, a state within the unstable region of the phase diagram had to be reached rapidly enough to suppress the ordinary nucleation-growth process. In the STM experiments with gold single crystals [5], this could be done by an electrochemical method. Through application of a microsecond voltage pulse to the STM tip, gold atoms from the topmost monolayer were randomly dissolved. The remaining Au atoms constituted a 2D lattice gas on a Au(111) surface. Because of attractive interactions between Au adatoms, the thus prepared statistically uniform state was however unstable and the process of its spinodal decomposition was taking place.

This process could be observed in situ in the same experimental setup by means of STM imaging. Figure 3.7 shows a typical pattern observed. Here, the arrow indicates the tip position where the voltage pulse was applied; the slow scanning direction is pointing downwards. Immediately after the pulse a labyrinthine pattern of monoatom-ically high Au islands appeared on the surface. It coarsened with time, increasing its characteristic length scale, as scanning proceeded.

Usually, phase separation leads to the development of macroscopic phase domains. If you take a sealed glass tube with vapor inside it and cool it, you will see, after waiting awhile, that all water droplets coalesce to form a layer of fluid at the bottom of the tube. The rest of it will be occupied by the vapor phase. There are however also systems where this does not take place and the so-called *micro-phase separation* is instead observed.

**Fig. 3.7** In situ STM image of the evolution of a labyrinthine, interconnected island pattern of Au adatoms on Au(111) surface. The slow scanning direction is pointed downwards. Reproduced from [5]

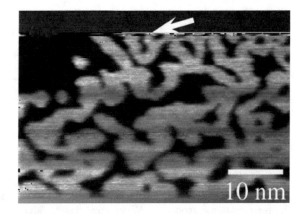

Let us again consider of mixture of two kinds of particles, A and B, with attractive interactions between particles A. Now, however, the particles will not be independent, but connected by links to form copolymers. First, long chains with regular alterations of particles A and B, such -A-B-A-B-A-A-B-, can be considered. In absence of links, this system would have undergone classical phase separation, with most of particles A eventually occupying a single macroscopic domain. When links are present this cannot however take place. Indeed, every particle A should always have at least two particles B—its neighbors in the chain—in its vicinity. Instead, copolymers condense into a compact state with regularly alternating layers of particles A and B [6].

Another situation where micro-phase separation occurs is encountered in systems where long-ranged repulsive interactions between particles are present in addition to short-ranged attractive interactions between them [7]. In this case, formation of macroscopic domains occupied by particles of the same kind is energetically non-favorable. If established, such large domains would tend to disintegrate due to repulsive forces between their parts. In surface systems, repulsive interactions can be due to the elastic strain that builds up in the substrate.

The phase diagram for systems with micro-phase separation is similar to that shown in Fig. 3.4. Inside the nucleation regions, sufficiently strong local perturbations are needed to initiate a growing domain of the new phase. The growth of a domain is however stopped when its size becomes comparable to the radius of repulsive interactions between the particles. Moreover, two phase domains cannot be separated by a distance shorter than such radius. Inside the spinodal decomposition region, the uniform state is unstable and its small perturbations with the form of plane waves are exponentially growing (or decaying) with time.

The characteristic dependence of the rate of growth on the wavenumber in the case of the microphase separation is shown in Fig. 3.8. In contrast to macroscopic phase separation, the growth of modes with small wavenumbers (and therefore large wavelengths) is now suppressed. The maximum rate of growth is found at some wavenumber $k_c$ and usually this wavenumber also determines the characteristic wavelength of the equilibrium spatially modulated phase.

**Fig. 3.8** Rates of growth of modes with different wavenumbers under micro-phase separation conditions

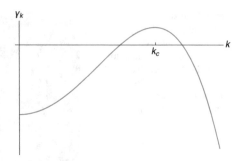

Complex patterns can develop during the phase transition process and some of them are common around us. When temperature falls below the freezing point, water vapor in the atmosphere condenses directly into the solid phase, i.e. into the ice. In the phase transition process, snowflakes are formed. Similar phenomena can occur on nanoscales. A particularly nice example is shown in Fig. 3.9 for the growth of graphene single crystals on copper foils affected by the presence of surface oxygen.

All kinetic processes are sensitively dependent on temperature. Whenever energy barriers need to be overcome in individual kinetic events, as, e.g., for hopping between the surface lattice sites, the Arrhenius law holds. This means that, below a certain temperature, the probabilities of transition become exponentially small and such transitions practically disappear. If, during a phase transition, the temperature drops down to such a point, the kinetic process becomes abruptly terminated and the transition pattern gets preserved as a metastable state. In winter, the flakes form snow that lies on the fields. Further transition from snow to ice is kinetically hindered (but it still can take place, so that, e.g., glaciers in the mountains become formed).

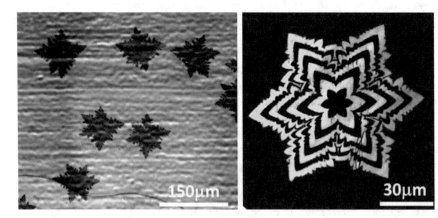

**Fig. 3.9** Growth of graphene single crystals on copper foils. The SEM image (*left*) and the respective isotope-labeled Raman map (*right*). Reproduced with permission from [8]

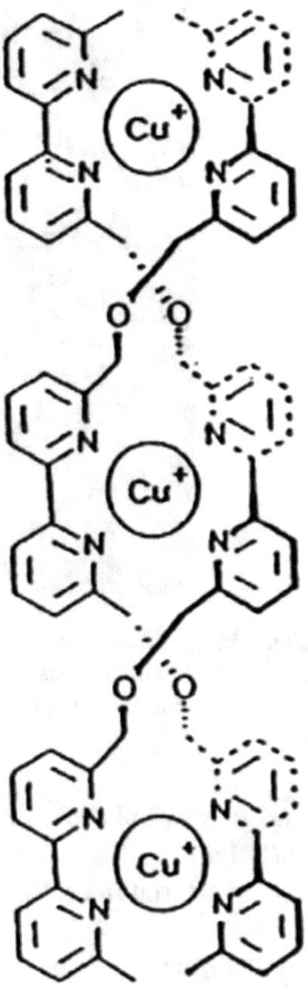

**Fig. 3.10** The copper double helix. From [10]

Beyond chemistry based on covalent bonds between atoms, the new field of supramolecular chemistry has evolved during the past decades [8]. Through interactions between molecules via non-covalent intermolecular forces, highly complex systems can be built through self-assembly. As an example, Fig. 3.10 shows a $Cu^+$ double helix formed by binding specific metal ions displaying tetrahedral coordination chemistry [9, 10]. By using metal coordination, hydrogen bonding or donor-acceptor interactions, a large variety of complex structures can thus be formed, with broad perspectives ranging from molecular recognition to molecular information processing.

The shapes developed through self-assembly processes may look very similar to those of biological organisms. Witherite is the material whose microscopic structure is formed by interwonen nanometer-sized rods of barium carbonate. The structure is characterized by nanoscale atomic ordering within the rods, but it lacks the long-ranged overall positional order. When such nanocrystalline aggregates grow, they do not express faces or edges characteristic for single crystals. Instead, various shapes bounded by smoothly curved surfaces and reminiscent of biological organisms are produced (Fig. 3.11). As pointed out by J. M. García-Ruiz et al. [11], such find-

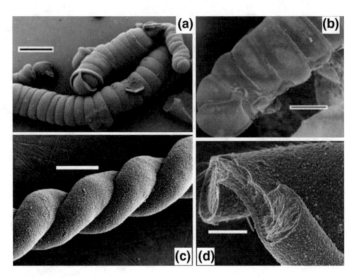

**Fig. 3.11** Field emission scanning electron microscopy (FESEM) images of mesoscopic inorganic filaments containing silica and witherite. *Scale bars* are 40 μm in (**a** and **b**), 10 μm in (**c**) and 4 μm in (**d**). Reproduced with permission from [11]

ings may even have implications for the search of remnants òf primitive life on the early Earth or Mars. The filaments in Fig. 3.11 look similar to the oldest terrestrial cyanobacterial microfossils believed to be more than 3 billion years old.

Whatever complex they may appear, self-assembled structure are however principally different from living biological organisms. Once a self-assembled state has been established, it can indefinitely persist in a closed system because it does not need to be nurtured by continuous free energy supply. Hence, it can also be described as the state with the "dead order".

All phenomena considered in this chapter are characteristic for a closed system permitting only heat exchange with the environment. The equilibrium state, as determined by the minimum of free energy, can often not be reached because of kinetic constraints. Thus, the full wealth of structures may develop, some examples of which have been shown above. In the literature, these phenomena are often denoted as "self-organization". We prefer instead the notion of "self-assembly" and reserve the term "self-organization" for open systems far from thermal equilibrium which will be the subject for the rest of the book.

# References

1. E. Schrödinger, Inaugural Lecture, University of Zürich. Original text: Die physikalische Forschung hat klipp und klar bewiesen, dass zumindest für die erdrückende Mehrheit der Erscheinungsabläufe, deren Regelmäßigkeit und Beständigkeit zur Aufstellung des Postulats der allgemeinen Kausalität geführt hat, die gemeinsame Wurzel der strengen Gesetzmäßigkeit der Zufall ist (1921)

2. J. Winterlinn, J. Trost, S. Renisch, R. Schuster, T. Zambelli, G. Ertl, Surf. Sci. **394**, 159 (1997)
3. J.W. Chan, J. Chem. Phys. **42**, 93 (1965)
4. D. Thron, Dissertation, Free University of Berlin, 2003
5. R. Schuster, D. Thron, M. Binetti, X. Xia, G. Ertl, Phys. Rev. Lett. **91**, 066101 (2003)
6. T. Ohta, K. Kawasaki, Macromolecules **19**, 2621 (1986)
7. M. Seul, D. Andelman, Science **267**, 476 (1997)
8. Y. Hao et al., Science **342**, 720 (2013)
9. J.M. Lehn, *Supramolecular Chemistry: Concepts and Perspectives* (VCH, Weinheim, 1995)
10. J.M. Lehn, Rep. Progr. Phys. **67**, 249 (2004)
11. J.M. García-Ruiz, S.T. Hyde, A.M. Carnerup, A.G. Christy, M.J. van Kranendonk, N.J. Welham, Science **302**, 1194 (2003)

# Chapter 4
# Self-organized Stationary Structures

In the literature the terms "self-assembly" and "self-organization" are frequently used as synonyms in order to describe processes by which a structure is formed spontaneously, in contrast to fabrication where every next step is controlled from outside. However, an important distinction between the two terms has to be made: Self-assembly concerns processes in thermodynamically closed systems with the tendency to reach an equilibrium, while self-organization is used for the development of ordered structures in open systems far from thermal equilibrium.

The essential difference consists in the fact that non-equilibrium self-organized structures require continuous supply of free energy for their existence in order to compensate for the ongoing formation of order and hence the loss of entropy. I. Prigogine proposed to denote such structures as *dissipative*, because they disappear once the inflow of free enery is terminated.

Another important aspect of self-organized structures is their kinetic origin. This means that their properties are governed by chemical reaction rates and transport processes, such as diffusion. For equilibrium structures, such dependence can never take place. Hence, for instance, if chemical reactions proceed within a closed system, they cannot affect the equilibrium states that become established inside it. In contrast to this, chemical reactions would often play a dominant role for dissipative structures that develop in open systems. While for self-assembly the kinetic processes are only of relevance during a transition to the state of thermal equilibrium, these are essential to stabilize the order in self-organizing systems.

The equilibrium self-assembled structures are static and cannot change if external parameters, such as temperature or pressure, are kept constant. In contrast to this, self-organized structures are *autonomous*, i.e. not completely controlled by their environment. They are intrinsically dynamic, so that temporal oscillations or propagating waves may develop without external time-dependent perturbations. It is the autonomy of self-organization structures and their active nature that make them ultimately responsible for the emergence of life.

© Springer International Publishing AG 2017                                                31
A.S. Mikhailov and G. Ertl, *Chemical Complexity*, The Frontiers Collection,
DOI 10.1007/978-3-319-57377-9_4

While thermodynamics of open systems sets the conditions under which self-organization takes place, it does not predict the mechanisms that govern it. The discovery of them has been a contribution of an entire generation of scientists and this research is still continued today. The pioneering theoretical research on the mechanisms of self-organization has been performed by the British scientist Alan Turing.

Turing was a mathematical genius and one of the founders of modern computer science. During the second world war he was successfully working on breaking military communication codes. Afterwards he moved to Manchester where the first operating British computer *Manchester Mark I* had been built for which he developed and maintained the software. The computer was mostly used in the program for the British atomic bomb test in 1952.

His strong interest in biology was arising from phyllotaxis, the arrangement of leaves on a plant stem or petals in a flower. The leaves form regular spiral patterns whose shape is closely related to the so-called Fibonacci numbers. Soon after his election in 1951 as a member of the Royal Society, Turing submitted for publication in this society's journal his fundamental paper *The chemical basis of morphogenesis* [1]. The abstract of this paper said: "It is suggested that a system of chemical substances, reacting together and diffusing through a tissue, is adequate to account for the main phenomena of morphogenesis.... The theory does not make any new hypotheses; it merely suggests that certain well-known physical laws are sufficient to account for many of the facts". Thus, Turing shared the view of von Bertalannfy, Schrödinger and Prigogine that there is no need to invent new laws for biological systems, the problem is how to apply the existing ones. It is not however clear whether he was familiar with their work.

In Turing's model two chemical substances, called *morphogens*, react one with another. The kinetics of the first substance $U$ is effectively autocatalytic; this chemical species enhances its own production. The reproduction of $U$ is controlled by the second chemical substance $V$ representing a product of $U$. When concentration of $V$ is increased, reproduction of species $U$ is slowed down. In a stirred flow reactor where fresh substrates are supplied and final products are evacuated, this chemical system has a steady state with some concentrations $u_0$ and $v_0$; this state is stable with respect to small perturbations.

As shown by Turing, the situation can radically change if both substances are distributed within a tissue and can diffuse, at different rates, within it. Provided that $V$ diffuses faster than $U$ and that its kinetics is also more rapid, the uniform state can become unstable with respect to periodic spatial modulation at some characteristic chemical wavelength $\lambda_0$. As a result of such *Turing instability*, a stationary dissipative structure develops in the tissue. The structure is characterized by the alteration of spatial regions where components $U$ or $V$ are predominantly expressed. Assuming that these chemical components control as morphogens the differentiation of biological cells, this would mean that a certain cellular structure has spontaneously developed starting from a uniform state.

The origin of such instability can be qualitatively rationalized. Suppose that, as a perturbation, a local region with enhanced concentration of the autocatalytic species $U$ became formed. Inside this region, more inhibitor $V$ will be produced. In absence of diffusion, this increase in the inhibitor concentration would have been enough to reduce the reproduction of activator species $U$ in such a way, that its concentration decreases and the steady state is recovered. But if the inhibitor species can diffuse, part of its additionally produced amount can spread however from the fluctuation region to the medium around it. Therefore, less inhibitor would be left inside the fluctuation and its concentration may be not enough to dampen the excessive local reproduction of the autocatalytic species $U$. On the other hand, more inhibitor would be found outside of the fluctuation and that would force the activator concentration in this region to decrease. Thus, the perturbation can further grow with time.

In mathematical terms, if chemical substances (morphogens) $U$ and $V$ are present and their local concentrations are $u$ and $v$, the considered equations of reaction and diffusion are

$$\frac{\partial u}{\partial t} = f(u, v) + D_u \frac{\partial^2 u}{\partial x^2} \tag{4.1}$$

$$\frac{\partial v}{\partial t} = g(u, v) + D_v \frac{\partial^2 v}{\partial x^2} \tag{4.2}$$

where functions $f$ and $g$ include all reaction terms, as well as supply and evacuation of the chemicals, and diffusion coefficients are $D_u$ and $D_v$. Equations (4.1) and (4.2) have a steady state where production and consumption of each substance are balanced; concentrations $u_0$ and $v_0$ in this steady state satisfy the equations $f(u_0, v_0) = 0$ and $g(u_0, v_0) = 0$.

Let us introduce perturbations $\delta u(x, t) = u(x, t) - u_0$ and $\delta v(x, t) = v(x, t) - v_0$. If they are small, Eqs. (4.1) and (4.2) can be linearized and we obtain

$$\frac{\partial \delta u}{\partial t} = a\delta u + b\delta v + D_u \frac{\partial^2 \delta u}{\partial x^2} \tag{4.3}$$

$$\frac{\partial \delta v}{\partial t} = c\delta u + d\delta v + D_v \frac{\partial^2 \delta v}{\partial x^2} \tag{4.4}$$

where coefficients $a, b, c, d$ are determined by the slopes of functions $f$ and $g$ in the steady state.

If species $U$ is (effectively) autocatalytic, the coefficient $a$ should be positive. If species $V$ represents an inhibitor, the coefficient $b$ should be negative. If $V$ is a (by)product of $U$, the coefficient $c$ is positive. Finally, since species $V$ is not autocatalytic, we have $d < 0$. Thus, conditions $a > 0, b < 0, c > 0$ and $d < 0$ will hold. In absence of diffusion, the steady state is stable if $a|d| < |b|c$.

Elementary perturbations of the uniform steady state are plane waves with different wavenumbers $k$ that can grow (or decay) with time at some rate $\gamma_k$. Therefore, $\delta u(x, t) \sim \exp(\gamma_k t) \sin[k(x - x_0)]$, $\delta v(x, t) \sim \exp(\gamma_k t) \sin[k(x - x_0) + \varphi_k]$. The rate of growth $\gamma_k$ and the spatial shift $\varphi_k$ are uniquely determined by the chemical system and the wavenumber $k$, whereas the wave position $x_0$ remains arbitrary.

Figure 4.1 shows the typical form of functions $\gamma_k$ near the instability threshold. Before the instability is reached, all perturbation modes decay with time, so that the entire dependence $\gamma_k$ lies below the horizontal axis. As the parameters are changed, the curve moves upwards and, at the instability threshold, it touches the horizontal axis at some point $k_0$. This means that the mode with the wavenumber $k_0$ is neither growing, nor decreasing exactly at the threshold. Above the threshold, this mode grows with time, together with a small group of modes with the wavenumber close to the critical wavenumber $k_0$.

The instability boundary in the parameter space has been determined by Turing [1]. Generally, the instability condition is that the ratio $D_v/D_u$ of the two diffusion constants should exceed some critical value, expressed as a function of the parameters $a, b, c$ and $d$.

The linear analysis can only yield the behavior of the system as long as it remains sufficiently close to the uniform state, so that the perturbations are small. At larger deviations from the initial uniform steady state, full nonlinear Eqs. (4.1) and (4.2)

**Fig. 4.1** The Turing instability. Rates of growth $\gamma_k$ of modes with different wavenumbers are shown

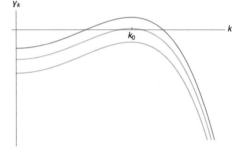

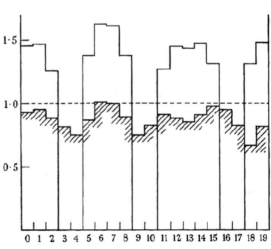

**Fig. 4.2** Original numerical simulation by Turing. The *dashed horizontal line* corresponds to the initial uniform state. The *dashed diagram* is an intermediate state. The final stationary state is shown by the *solid line*. Reproduced with permission from [1]

have to be used. As already noted by Turing, two situations are possible. It can be that the nonlinearities further accelerate the growth of the first unstable modes. In this case (which corresponds to a *subcritical bifurcation* in mathematical terms), the final structure may well be very different from the first unstable mode (and oscillations or waves may even develop). It is however also possible (*a supercritical bifurcation*) that nonlinear terms dampen the critical mode and lead to the saturation of its growth while the amplitude of the mode is still small. Under the latter conditions, the properties of the final stationary structure can be well understood by considering only the critical mode given by the investigation of the linearized system.

While the linear stability analysis applies for any system and therefore its results are universal, the nonlinear evolution and the final state depend sensitively on the specific form of the functions $f(u, v)$ and $g(u, v)$ in Eqs. (4.1) and (4.2). Generally, only numerical integration of the evolution equations governing self-organized structures is possible. Therefore, advances in understanding of self-organization phenomena have been closely related to the progress in computer technology.

As already noted, Turing was responsible for the development of software and running of simulations on *Manchester Mark I*. While most of the simulations should have been dealing with military applications, Turing had managed to use this computer in his own research. Thus, first numerical simulations of nonlinear self-organization phenomena were performed. Figure 4.2 shows an example of a stationary pattern arising from the uniform initial state. The numerical integration was carried out for a set of hypothetical chemical reactions in a system that represented a ring of 20 diffusively coupled cells. The dashed diagram shows an intermediate state and the solid line corresponds to the final stationary structure. The intermediate state looks irregular because two modes with different periods were simultaneously unstable. The instability in the considered system was "catastrophic", or corresponding to a subcritical bifurcation in modern terms. Therefore, the final state was very different from the early intermediate state.

For comparison, we show in Fig. 4.3 stationary Turing patterns obtained by numerical simulation on a modern computer [2]. The model parameters were tuned to the supercritical case, so that the final stationary pattern (Fig. 4.3a) was similar to that of the first unstable mode. This is clearly seen in Fig. 4.3b where temporal evolution starting from the initial uniform state is displayed. Here, the spatial coordinate is plotted along the horizontal axis and time corresponds to the vertical axis. The local concentration of the autocatalytic species is shown by color coding, with the highest concentration represented by the red color. After the pattern has emerged, only a small rearrangement of spikes takes place until the final state is reached. The last two panels in Fig. 4.3 show an example of a Turing pattern in the two-dimensional medium. The autocatalytic activator species (Fig. 4.3c) is localized within narrow spots that form a regular lattice. The inhibitor species (Fig. 4.3d) is depleted within the spots, but its concentration is enhanced around them.

Although it may seem paradoxical, emergence of order implies that the symmetry of a system gets decreased. Following Turing, let us consider as an example morphogenesis in a tissue ring. The initial uniform non-differentiated state is obviously symmetric with respect to all shifts (or *translations*). If, as a result of an instability,

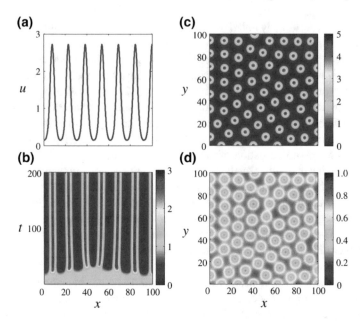

**Fig. 4.3** Modern numerical simulations of Turing patterns. Final stationary pattern (**a**) and the space-time plot (**b**) showing its development from the initial unstable uniform state. Distributions of activator (**c**) and inhibitor (**d**) in the final stationary Turing pattern in the two-dimensional medium. *Color coding* of concentration variables according to the *bars* displayed on the *right sides* of the plots. Reproduced with permission from [2]

a periodically varying spatial pattern has developed, only the symmetry with respect to the shifts by the spatial period however remains. Thus, the overall symmetry of the system becomes reduced.

Such *spontaneous symmetry breaking* may also appear paradoxical from the mathematical point of view. Indeed, if the system described by Eqs. (4.1) and (4.2) was initially in the uniform stationary state, such that $f(u_0, v_0) = 0$ and $g(u_0, v_0) = 0$, then it will continue indefinitely to stay in it. Here, it has to be noted that small perturbations are always present in a physical system. Such initial local deviations from the uniform state are very important because, if this state is unstable, they can grow with time, so that a transition to the new, less symmetric, stationary state takes place.

According to Turing, "the situation is very similar to what arises in connexion with electrical oscillators. It is usually easy to understand how an oscillator keeps on going when once it has started, but on a first acquaintance it is not obvious how the oscillation begins. The explanation is that there are always random disturbances present in the circuit. Any disturbance whose frequency is the natural frequency of the oscillator will tend to set it going. The ultimate fate of the system will be a state of oscillation at its appropriate frequency, and with an amplitude (and the wave form) which are also determined by the circuit. The phase of the oscillation alone is determined by the disturbance".

For periodic stationary self-organized structures, this means that only the positions of wave maxima in space (corresponding to the oscillation phase) are determined by small initial perturbations, whereas the spatial period, the amplitude and the profile of the stationary pattern are uniquely determined by the system. Hence, the dependence on the initial conditions is strongly limited: they can affect only one aspect of a self-organized structure, i.e., its position in space. Other aspects of the structure, i.e., its amplitude and shape, are not influenced by them.

While they were discovered by Turing in the framework of biological morphogenesis, these properties are general and characteristic for self-organized structures of any origin. In a fact, such structures are considered as being *self* organized exactly because, to a large extent, they are determined and controlled by the system itself.

Another important issue is the response of a self-organized structure to external perturbations. Similar to self-oscillations in electrical circuits, such perturbations have only a weak effect on the shape or the amplitude of the developed structures. Indeed, the system tends to keep such properties fixed and therefore the deviations are suppressed. On the other hand, the position of the pattern in space, which is arbitrary and can be also influenced by the initial conditions, can be easily changed when perturbations are applied. It is because of a limited response to external perturbations that self-organized systems can be viewed as *autonomous*, i.e. not completely controlled by their environment.

The concept of spontaneous symmetry breaking arises also in the theory of equilibrium second-order phase transitions in physical systems developed by Lev Landau in 1937 [3]. Consider, for example, a transition from the paramagnetic to the ferromagnetic state in a system of interacting atomic spins that takes place when temperature is decreased. In the high-temperature paramagnetic state, the net magnetization is zero and therefore this state is symmetric with respect to any rotation. On the other hand, the low-temperature ferromagnetic state is characterized by a certain magnetization vector. Hence, it remains symmetric only with respect to rotations around the magnetization vector and the symmetry of the system becomes reduced (or *broken*). Again, the direction of the final magnetization vector after the transition is determined by small perturbations in the initial conditions.

Landau has proposed [3] a description of the systems near the phase transition where the principal role is played by the *order parameters* specifying the degree of symmetry breaking. Similar descriptions are possible and have been developed for self-organization phenomena in the systems far from thermal equilibrium. The notion of order parameters for such systems has been introduced and explored by Haken, with the applications ranging from physics, chemistry and biology to sociology and economics [4]. Some of the models used in the theory of self-organization phenomena represent a direct extension of the theoretical descriptions by Landau (as discussed later in the book).

Returning to the mathematical problem studied by Turing, several comments can be made. First, it should be noted that the instability is also possible if the second species $V$ does not inhibit the reproduction of the autocatalytic species $U$, but acts as a substrate (or "fuel") for such reproduction. In terms of the linearized Eqs. (4.3) and (4.4), this means that the conditions $b > 0$ and $c < 0$ are satisfied, i.e. that an increase in the substrate concentration enhances production of $U$ and that the species $V$ is consumed by $U$. The existence of instability in this case follows from the fact that by transformation $\delta v \rightarrow -\delta v$, we return to the Eqs. (4.3) and (4.4) with $b \rightarrow -b$ and $c \rightarrow -c$ and thus we return to the previously considered case. If the concentration of the autocatalytic species is locally increased, this leads to a depletion of the substrate not in the perturbation, but, due to rapid diffusion of the substrate, also in the region around it, thus suppressing the reproduction of $U$ in the surrounding area.

Second, the instability can furthermore be found in systems with three or more components, although the analysis becomes more complicated in this case. It is not however possible if only one species is present. Note that, as already proven by Turing [1], only stationary spatial patterns can develop in chemical systems with two components. In contrast, three-component models can additionally support a different kind of instability leading to spontaneous development of traveling waves. This latter instability is known as the "wave bifurcation".

Third, comparing Figs. 4.1 and 3.8, one can notice a similarity between the Turing instability and the phase transition leading to micro-phase separation in equilibrium physical systems. This similarity is not accidental and the mathematical descriptions of these two phenomena are indeed close. The principal difference is that the micro-phase separation is an equilibrium transition and therefore the conditions for it, such as the critical transition point, are determined by thermodynamic properties of a system. On the other hand, the Turing instability is of kinetic origin and it is determined by the kinetic coefficients, such as diffusion constants and chemical reaction rates.

In his pioneering publication, Turing not only considered a general mathematical problem, but also proposed two hypothetical systems of irreversible chemical reactions where the instability could be observed. His first reaction system was

$$X + Y \rightarrow W, W + A \rightarrow 2Y + B, 2X \rightarrow W, A \rightarrow X, Y \rightarrow B, Y + C \rightarrow C^* \rightarrow X + C.$$
$$(4.5)$$

The concentration of substrate $A$ was kept constant and the catalyst $C$ (together with its form $C^*$) was uniformly distributed in the medium. Some of the reaction steps were fast, so that the intermediate product $W$ could be adiabatically eliminated. Hence, the system could be described by a set of two kinetic equations for concentrations of species $X$ and $Y$ that had the form of Eqs. (4.1) and (4.2). This model was numerically integrated on the *Manchester Mark I* computer and the results shown in Fig. 4.2 could thus be obtained.

This system was however characterized by Turing as "somewhat artificial" (perhaps, because it involved a trimolecular reaction step). Therefore, another example was also presented by him where all reactions were bimolecular. It described conversion of substrates $A$ and $B$ into final products $D$ and $E$ proceeding through reaction steps involving intermediate products $X$, $Y$, $C$, $V$, $V^*$ and $W$. Explicitly, this scheme had the form

$$A \rightarrow X, X + Y \rightleftarrows C, C \rightarrow D, B + C \rightarrow W \rightarrow Y + C, Y \rightarrow E, Y + V \rightarrow V^* \rightarrow E + V \tag{4.6}$$

The concentrations of substrates were maintained constant. Moreover, it was assumed that some reaction steps were fast and intermediate products $C$, $V$, $V^*$ and $W$ could therefore be adiabatically eliminated. As a result, only a system of kinetic equations for concentrations of two species, $X$ and $Y$, was left. Note that this reaction scheme does not explicitly include an autocatalytic step. There is however a positive feedback loop inside it.

In 1967, Grégoire Nicolis and Ilya Prigogine [5] have performed the analysis for the second reaction system in terms of nonequilibrium thermodynamics. A year later, Réne Lefever and Ilya Prigogine [6] have proposed three further model systems, including a model with only two intermediate species,

$$A \rightarrow X, 2X + Y \rightarrow 3X, B + X \rightarrow D + Y, X \rightarrow E. \tag{4.7}$$

It described conversion of substrates $A$ and $B$, present at constant concentrations, into the products $D$ and $E$. The authors have also remarked that the scheme was "physically unrealistic" because of the trimolecular step. Nonetheless, the model (4.7) became popular because of its simplicity; it is known as the *Brusselator*. Many numerical simulations of Turing patterns have been performed using it.

Independently, Alfred Gierer and Hans Meinhardt developed in 1972 a mathematical model [7] for biological pattern formation; they learned about the work of Turing only from the referee, after submitting their paper to a journal. Their model was formulated in terms of two components, $a$ and $h$. It was phenomeno-logical and did not correspond to a set of chemical reactions. The equations of the Gierer–Meinhardt model are

$$\frac{\partial a}{\partial t} = \rho \frac{a^2}{h} - \mu_a a + D_a \frac{\partial^2 a}{\partial x^2} + \rho_a \tag{4.8}$$

$$\frac{\partial h}{\partial t} = \rho a^2 - \mu_h h + D_h \frac{\partial^2 h}{\partial x^2} + \varrho_h \tag{4.9}$$

where $\rho$, $\rho_a$ and $\rho_h$ are constant parameters. The component $a$ was described as an *activator* and the component $h$ as an *inhibitor*. Note that, although we have already used these two terms above, they were first proposed by Gierer and Meinhardt.

Despite the fact that Turing patterns were already extensively theoretically dis-cussed, their experimental observation for chemical systems had to wait for another twenty years. The reason was that the theory requires that the inhibitor diffuses much faster than the activator. But, in aqueous solutions, diffusion coefficients of various species are not so different from each other. The success was attained more or less by an accident: In an oscillating reaction in which the autocatalytic positive feedback was controlled by iodine ions, these could bind to the immobile starch used as a color indicator. The formation of this complex slowed down the diffusion of the activator considerably.

The first observation of Turing patterns has been performed in 1990 by Patrick De Kepper with coworkers in Bordeaux [8, 9]. In the experiments, thin strips of gel ($20 \times 3 \times 1$ mm) were used to prevent convection. The selected chemical system was the chlorite-iodide reaction with the added malonic acid. To make concentration changes visible (and originally just for that purpose!), the gel was loaded with a solvable starchlike color indicator. The opposite long edges of the stripe were in contact with two well mixed reservoirs where different reactants were present. The reactants were diffusing from opposite sides into the slab and, in the center of it where both reactants were present, the reaction could proceed.

The Turing instability was manifested in the spontaneous formation of a line of spots along the stripe. Their wavelength $\lambda = 0.2$ mm was much smaller than the geometric sizes of the reactor and seemed to be intrinsic. The pattern could be sustained indefinitely and persisted unchanged for more than 20 h. Most importantly, the pattern reappeared with the same wavelength after it was destroyed by a temporary perturbation.

A year later, the experiments were repeated by Qi Ouyang and Harry L. Swinney in Texas [9] under well-controlled conditions and in the geometric set-up where observation of (quasi) two-dimensional patterns was possible. Instead of a stripe, a thin sheet of gel squeezed between two well-mixed reservoirs was used. The reactants were supplied uniformly from the reservoirs and the reaction could proceed in the middle plane. The chemical system was the same as before and starch as the color indicator was also used.

The patterns emerged spontaneously from the initial uniform background. First, many circles were formed, but, after an hour, they broke into small spots that evolved more slowly. Eventually, the spots formed a hexagonal array with some grain boundaries. The boundaries were moving very slowly, by about only 0.2 mm per day. At high iodide or low malonic acid concentrations, stripes were formed instead of the spots. Furthermore, the transition from the uniform state to a hexagonal pattern was investigated by using temperature as the control parameter. The critical instability point was at about 18 C and, at lower temperatures, Turing patterns could be observed. The amplitude of the patterns was gradually decreasing and vanishing at the instability point, so that the supercritical (or a weakly subcritical) bifurcation was apparently responsible for them.

Subsequently, various designs for the observation of Turing patterns in chemical reactions have been explored [10]. To illustrate the experimentally observed Turing patterns, we have chosen an example [10] where a hydrogen-ion autoactivated reaction, the thiourea-iodite-sulphite (TuIS) reaction, has been employed. These experiments were performed with the one-side-fed unstirred spatial reactor representing a gel disc. Figure 4.4 shows different kinds of patterns that could be found by varying the reactants concentrations. The development of a Turing pattern of spots from the initial uniform state is illustrated in Fig. 4.5. In Fig. 4.5b, the concentration along the central linear cross-section is plotted versus time. Spontaneous emergence of the pattern and the saturation of its growth are clearly seen (compare Fig. 4.3b).

An alternative method that allows to control diffusion of selected chemical components consists of working with microemulsions [11]. They are formed by nanodroplets of water surrounded by surfactant molecules and residing in the continuous oil phase. If a reaction involves both water-soluble and oil-soluble components, the water-soluble species are confined within the nanodroplets and can only diffuse slowly together with them. On the other hand, small oil-soluble molecules can diffuse through the oil at typical diffusion rates. In the experiments, the Belousov-Zhabotinsky reaction (see Chap. 7) was used and a component acting as an inhibitor in this reaction was oil-solvable. Thus, a system with a fast diffusing inhibitor could be constructed and Turing patterns have been observed (see [11]).

A variant of the Turing mechanism was realized with an electrochemical system at an electrode surface [12]. The role of a rapidly diffusing inhibitor species was then played by long-ranged electrostatic interactions. Surface migration of ions under the influence of the electric field was replacing the diffusion in this case and Turing-like patterns could thus be seen [12].

The situation in biological morphogenesis has turned out to be more complex than assumed by Turing. For him, the genes were merely catalysts and therefore they were not included into the model construction. Now we know that the activity of genes

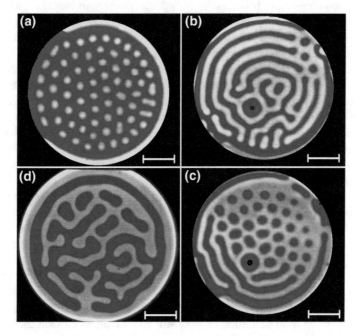

**Fig. 4.4** Turing patterns in the TuIS reaction. **a** Hexagonal array of spots, **b** stripe pattern, **c** coexistence of spots and stripes, **d** irregular branch pattern. *Scale bars* equal 4 mm. Reproduced with permission from [10]

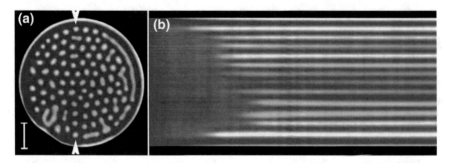

**Fig. 4.5** Development of a Turing pattern from the initial uniform state. Evolution of the concentration distribution along the central vertical cross-section, indicated by *arrows* in **a**, is displayed in the space-time plot (**b**) where time runs along the *horizontal axis*. Reproduced with permission from [10]

in a cell is highly regulated, so that a level of expression of a given gene sensitively depends on the state of activity of other genes in a network. While proteins are expressed by genes, the involved mechanisms are complex and cannot be reduced to a set of chemical reactions, as assumed by Turing. Moreover, most of the expressed proteins are confined within a cell and cannot act as morphogens that are secreted into the extracellular space and diffuse over long distances. It may well be that, instead of diffusion, communication between the cells proceeds through elastic deformations. Such deformations develop when the cells are in contact one with another and they change their shape (Turing knew about such deformations but neglected them in his theory). Finally, it is not correct that, at the beginning of the process, spherical symmetry is present and then its spontaneous breakdown occurs. When an egg is fertilised, the position where the spermatozoid has entered the egg is known to play an important role in the subsequent development of the fetus, anchoring its orientation. Therefore, the theory of Turing is not currently seen as providing a fundamental general explanation of biological morphogenesis. Nonetheless, there are also studies which suggest that the Turing mechanism is in operation in actual pattern formation processes in biological organisms. A discussion of the respective biological aspects is however beyond the scope of this book.

Returning to mathematics, one extension of the original theory has to be mentioned. According to Turing, pattern formation takes place when the uniform stationary state is unstable. It is accompanied by spontaneous symmetry breaking and exponential enhancement of weak perturbations present in the unstable initial state. As already noted above, this is similar to the situation with second-order phase transitions in physical systems at thermal equilibrium. It is however also possible that the uniform stationary state remains stable with respect to small perturbations, but self-organized patterns nonetheless develop. Namely, the instability with respect to sufficiently strong local perturbations may still be present: when such perturbations are applied, a transition to the self-organized structure is triggered. In equilibrium physical systems, this behavior is characteristic for the *first-order* phase transitions where a critical nucleus needs to be created in order that the transition occurs.

If the instability is "catastrophal" in Turing's words (or *subcritical* in modern terms), it becomes accelerated when deviations from the uniform stationary state increase. For subcritical instabilities, hysteresis behavior is however characteristic. This means that, if we start from the self-organized pattern above the critical point and gradually decrease the control parameter going into the region where the uniform stationary state is stable, the pattern will persist below the critical point. Thus, there will be a parameter region where the self-organized pattern coexists with the uniform stationary state. In this region, the uniform state is stable with respect to small perturbations. However, if a strong enough perturbation—a *critical nucleus*—is created, a transition to the self-organized structure would take place.

Slightly below the critical point, the coexisting spatial pattern would be similar to that found above it, retaining the periodicity and occupying the entire medium. Farther away from such point, qualitative differences may however arise. Particularly, *localized* structures can then develop that would represent a group of spots (or even a single spot) surrounded by the medium in the uniform stationary state.

Suppose that, through a perturbation, a small region (a spot) with the increased activator concentration has been formed (Fig. 4.6). Inside this region, the inhibitor would be excessively produced. Because this species is rapidly diffusing, the inhibitor would have to go outside of the activator spot and form a cloud surrounding it. Within this cloud, expression of the activator would be suppressed, so that the activator spot cannot expand. Thus, a localized stable stationary structure, a *hot spot*, can arise. In a fact, it would look like one of the spots in the hexagonal arrays in Fig. 4.3 or 4.4.

Despite a close relationship between Turing patterns and localized dissipative structures in activator-inhibitor systems, the theory of such structures had to be independently developed. The reason was that such structures are essentially nonlinear and therefore they cannot be found within the linear approximation of Turing's theory. This further implies that the conditions for the existence of localized structures, as well as their properties, should strongly depend on the form of the nonlinearity—and thus on a specific reaction-diffusion model. Consequently, no universal theory of localized structures in activator-inhibitor systems can exist.

**Fig. 4.6** Stable localized structure (a spot). Activator (*full line*) and inhibitor (*dashed line*) concentration distributions are shown. From [13], modified

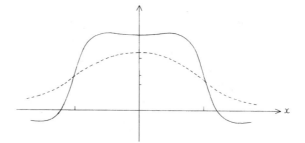

Nonetheless, principal properties of such structures could be explored through investigation of characteristic models. One of the first studies where analytical solutions for localized structures were constructed was reported [13] by S. Koga and Y. Kuramoto in 1980. These authors have also considered localized structures in bistable media where they can represent broad domains. The stability of the spots was analyzed and spontaneous development of periodic breathing in such localized structures was considered. The respective solutions for two-dimensional media were constructed and their stability was investigated by T. Ohta, M. Mimura and R. Kobayashi [14]; see also the publication [15] where an instability leading to traveling spots has been pointed out.

In addition to treating the tissue as a continuous medium, Turing has also considered a model with a ring of cells that were diffusively coupled. The discrete models are more appropriate for an early stage of the embryo development when the number of cells is relatively small. The cells form a cluster and there are physical contacts between their membranes. It is therefore possible that some molecules are exchanged directly between the neighbouring cells, i.e. without their release into the extracellular space. Hence, the models with discrete cells and diffusive coupling between them are indeed appropriate.

However, the shape of an early embryo is different from a linear ring. When physical contacts between the cells in an embryo are examined, a complex network of interconnected cells can be constructed (see, e.g., [16]). Thus, a more appropriate model would consider a network of coupled chemical microreactors. The reactions proceed inside the reactors and some chemicals can be transported diffusively between the reactors if there is a link between them in the network.

Already in 1971, H. G. Othmer and L. E. Scriven [17] have extended the mathematical theory to arbitrary networks of diffusively coupled reactors. The most important difference was that the critical modes, that start to exponentially grow above the instability point, did not represent plane waves. The form of such modes was more complex and sensitive to the topology of connections in a given network. In mathematical terms, they represented eigenvectors of the Laplacian matrix that is determined by the pattern of connections between the nodes. While the theory was general, only examples of regular lattices were however explicitly considered, for which the Laplacian eigenvectors reduce to plane waves.

For three decades, the network problem remained not again addressed. In 2004, W. Horsthemke, K. Lam and P. K. Moore [18] have published an analysis for small arrays of diffusively coupled reactors. More recently, an extensive statistical study for large random networks has been performed [19]. It has been shown that critical Turing modes are localized on definite subsets of network nodes. Numerical simulations have revealed that the instability is always "catastrophal" (or subcritical), so that the final established pattern differs much from the structure implied by the critical mode. Moreover, a strong hysteresis is typically observed and many different self-organized stationary structures can coexist.

# References

1. A.M. Turing, Phil. Trans. R. Soc. Lond. B **237**, 37 (1952)
2. P.K. Maini, T.E. Wooley, R.E. Baker, E.A. Gaffney, S.S. Lee, Interface Focus **6**, 487 (2012)
3. L.D. Landau, Zh. Eksp. Teor. Fiz. **7**, 19 (1937)
4. H. Haken, *Synergetics. Nonequilibrium Phase Transitions and Self-Organization in Physics, Chemistry and Biology* (Springer, Berlin, 1977)
5. I. Prigogine, G. Nicolis, J. Chem. Phys. **46**, 3542 (1967)
6. I. Prigogine, R. Lefever, J. Chem. Phys. **48**, 1695 (1968)
7. A. Gierer, H. Meinhardt, Kybernetik **12**, 30 (1972)
8. V. Castets, E. Dulos, J. Boissonade, P. De Kepper, Phys. Rev. Lett. **64**, 2953 (1990)
9. Q. Ouyang, H.L. Swinney, Nature **352**, 610 (1991)
10. J. Horvath, I. Szalai, P. De Kepper, Science **324**, 772 (2009)
11. V.K. Vanag, I.R. Epstein, Chaos **18**, 026107 (2008)
12. Y.-J. Li, J. Oslonovich, N. Mazouz, K. Krischer, G. Ertl, Science **291**, 2395 (2001)
13. S. Koga, Y. Kuramoto, Progr. Theor. Phys. **63**, 106 (1980)
14. T. Ohta, M. Mimura, R. Kobayashi, Physica D **34**, 115 (1989)
15. K. Krischer, A.S. Mikhailov, Phys. Rev. Lett. **73**, 3165 (1994)
16. F.A. Bignone, J. Biol. Phys. **27**, 257 (2001)
17. H.G. Othmer, L.E. Scriven, J. Theor. Biol. **32**, 507 (1971)
18. W. Horsthemke, K. Lam, P.K. Moore, Phys. Lett. A **328**, 444 (2004)
19. H. Nakao, A.S. Mikhailov, Nat. Phys. **6**, 544 (2010)

# Chapter 5
# Chemical Oscillations

Oscillations, i.e. processes that vary periodically with time, are common in living systems; they are characteristic for breathing or the heart beat. There are however also purely chemical systems where these phenomena are observed. Note that, since at thermal equilibrium all systems are in stationary states, sustained oscillations can only take place in open systems far from equilibrium. Although systematic studies of oscillation effects for chemical reactions started only in the second half of the twentieth century, there were also earlier reports.

The German chemist Wilhelm Ostwald, one of the founders of physical chemistry, studied in 1899 electrochemical dissolution of chromium in acids. It had been observed [1] that, at certain applied voltages, the metallic sample could be in one of two states: It was either undergoing rapid dissolution accompanied by evolution of hydrogen and the formation of salt, or it remained inert (passive). Usually chromium was passive in diluted acid. Warming up caused, however, a transition to the active state with intense oxidation.

Detailed experiments were performed where the rate of gas evolution was recorded. For this purpose the apparatus shown in Fig. 5.1 was developed: Upon production of gas, the pressure increased in a thin tube and its variations were recorded by a pen drawing a trace on an unwinding roll of paper so that registration became fully automatized.

An example of the experimental data is shown in Fig. 5.2. Each oscillation starts with a sudden increase of the reaction rate, followed by a slow decrease after passing a maximum. The oscillation period increases slowly with time while the amplitude decreases, presumably because of the consumption of the acid. Oscillations with more complex waveforms and irregularities as well as similar effects with sulphuric acid were also observed, and also the influence of temperature and other parameters was studied. However, the reproducibility was poor, and Ostwald concluded that he refrains from proposing any possible mechanism until these measurements become clearly reproducible on the basis of well-defined prescriptions [1].

© Springer International Publishing AG 2017
A.S. Mikhailov and G. Ertl, *Chemical Complexity*, The Frontiers Collection,
DOI 10.1007/978-3-319-57377-9_5

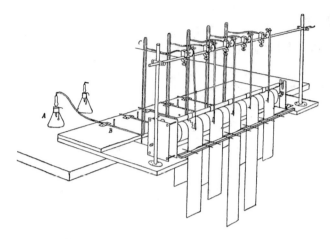

**Fig. 5.1** Experimental apparatus used by Ostwald to study oscillations of the rate of electrochemical dissolution of chromium. Reproduced from [1]

A first attempt for theoretical explanation for chemical oscillations was made [2] in 1910 by the American Alfred Lotka, who was at that time a PhD student in Birmingham. In his model two substances $A$ (the substrate) and $B$ (the product) were interacting with each other in a way that $B$ was autocatalytically affecting its production from $A$. The substrate was provided with a constant rate $H$ and the product was formed in a bimolecular reaction with $A$ and removed by a decay reaction. The corresponding kinetic equations read as

$$\frac{dc_A}{dt} = H - kc_A c_B \tag{5.1}$$

$$\frac{dc_B}{dt} = kc_A c_B - k_2 c_B \tag{5.2}$$

The solutions of this model yielded, however, only damped oscillations. Sustained oscillations were obtained in the modified model [3] where the supply of $A$ was governed by its own concentration, so that the following kinetic equations resulted

$$\frac{dc_A}{dt} = k_1 c_A - kc_A c_B \tag{5.3}$$

$$\frac{dc_B}{dt} = kc_A c_B - k_2 c_B \tag{5.4}$$

The same equations were derived independently by the Italian Vito Volterra in the book *Leçons sur la Theorie Mathématique de la Lutte pour la Vie* [4]. In the context of ecology, species $A$ and $B$ could be considered as prey and predator, respectively.

**Fig. 5.2** Rate oscillations during electrochemical dissolution of chromium in hydrochloric acid. Subsequent time spans of one-hour duration are shown. Reproduced from [1]

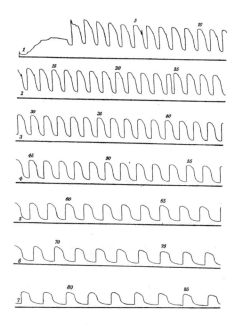

Equation (5.3) then described the reproduction of prey $A$ and its consumption by the predators $B$, while Eq. (5.4) modeled the reproduction of predators $B$ by feeding on the pray $A$ and their disappearance by natural death. Nowadays, these equations are known as the *Lotka-Volterra model*.

A typical solution of the Lotka–Volterra model, shown in Fig. 5.3, exhibits temporal oscillations of both variables, with the peaks in the numbers of predators lagging behind the peaks of the prey. Remarkably, this agrees qualitatively with the reported annual numbers of furs from hares and lynxes delivered to the Hudson's Bay Company in the period from 1845 to 1935, as shown in Fig. 5.4.

As Lotka remarked [2], "no reaction is known that follows the above law and as a matter of fact the case here considered was suggested by the consideration of matters lying outside the field of physical chemistry". However, we will later demonstrate that chemical systems exhibiting oscillatory kinetics can indeed be described by nonlinear equations of this type.

Returning back to chemical oscillations: While the study of Ostwald concerned a heterogeneous reaction involving the interface of a solid, oscillatory effects in the kinetics of a homogeneous reaction in solution were first reported [6] in 1921 by William Bray who studied the combined oxidation of iodine to iodic acid and the reverse reduction of iodic acid to iodine, i.e.

$$5H_2O_2 + I_2 \rightarrow 2HIO_2 + 4H_2O \tag{5.5}$$

$$5H_2O_2 + 2HIO_2 \rightarrow 5O_2 + I_2 + 6H_2O. \tag{5.6}$$

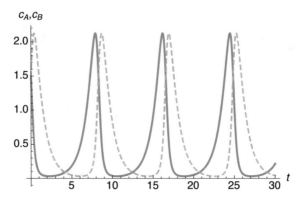

**Fig. 5.3** Sustained oscillations in the concentrations of species $A$ (prey, *solid line*) and $B$ (predators, *dashed line*) in the Lotka–Volterra model. The parameters are $k = k_1 = k_2 = 1$

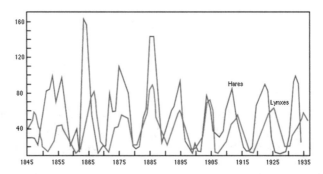

**Fig. 5.4** Annual variation in the numbers of furs (in thousands) of hares and lynxes delivered to the Hudson's Bay Company. Redrawn with the data from [5]

Here, hydrogen peroxide plays a dual role as an oxidizing and reducing agent. Remarkably, the iodic acid formed in the first reaction (which is the fast one) acts as a substrate in the slow second reaction. Thus this reaction accelerates as further hydrogen peroxide becomes converted, reflecting an autocatalytic character.

Although iodine was consumed in reaction (5.5) and was then again produced in reaction (5.6), its concentration did not necessarily remain constant, but slowly oscillated under certain conditions in parallel to the evolution of oxygen. At 60 °C, the period was of the order of several minutes and decreased to about two days when the temperature was lowered to 25 °C, where the oscillations persisted for about a month (Fig. 5.5). Their mechanism remained unclear, but Bray suggested some relation to the Lotka model because of the autocatalytic effect.

Oscillations can readily be found in the theoretical activator-inhibitor models discussed in the previous chapter. When Turing was analyzing the stability of the uniform steady state in the framework of Eqs. (4.4) and (4.5), he showed that, depending on the parameters, uniform oscillations could develop as well [7]. He also remarked that "there are probably many examples of this metabolic oscillation, but no really satisfactory one is known to the author".

One of the principal metabolic processes taking place in all living cells is glycolysis. Here, molecules of ATP (adenosine triphosphate), serving to supply energy to active proteins inside the cell, are synthesized from glucose. Thus when wine is produced from grape juice the glycolysis in yeast is responsible for fermentation, whereby ethanol is formed as a byproduct in a series of reactions catalyzed by enzymes. Identification of the glycolysis pathway by Otto Meyerhof was one of the major biochemical discoveries of the twentieth century.

Britton Chance developed an optical method to monitor the progress of metabolic processes in vivo. Later, this method was modified by L. Duysens and J. Amesz who observed the development of a few damped oscillations following a perturbation [8]. Sustained glycolytic oscillations in cell suspensions were then studied in detail by Chance and were found to be controlled by the enzyme phosphorfructokinase [9]. This enzyme is allosterically activated by its product, thus providing a positive feedback loop which is essential for the development of oscillations. Subsequently, Benno Hess became involved in this research, who could reproduce the glycolysis oscillations in the cell-free extract of yeast thus confirming that these were of purely chemical origin, with the substrate trehalose being essential for them [10]. An example of the observed oscillations that could extend up to 22 h is given in Fig. 5.6.

In subsequent chapters, two systems exhibiting chemical oscillations will be described in detail, namely the Belousov-Zhabotinsky system as a homogeneous reaction which serves as a kind of "drosophila" for the whole field (Chap. 7) and heterogeneous catalytic reactions at surfaces (Chap. 8). In the rest of this chapter, the focus is mainly on theoretical aspects.

The Lotka–Volterra model described by Eqs. (5.1) and (5.2) has provided inspiration for many studies, both in chemistry and biology. Nonetheless, this system is special in the mathematical sense. It can be checked that it possesses a conserved

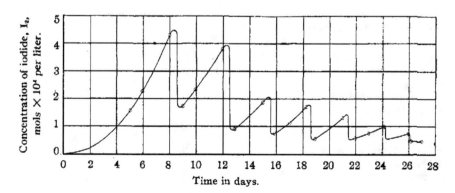

**Fig. 5.5** Dependence of iodine concentration on time (in *days*). Reproduced with permission from [6]

quantity $Q = -kc_A + k_2 \ln c_A - kc_B + k_1 \ln c_B$, such that $dQ/dt = 0$. This quantity is determined by initial conditions and defines a curve on the plane of concentrations $c_A$ and $c_B$ along which the dynamics proceeds. Thus, different initial conditions give rise to oscillations with different amplitude, frequency and shape (see Fig. 5.7). Such sensitivity to initial conditions is a unique feature of this particular model - a different kind of behavior is generally observed.

As a characteristic example, the Brusselator model corresponding to the reaction scheme (4.8) can be chosen. Its kinetic equations for concentrations of chemical species $X$ and $Y$ are

$$\frac{dc_X}{dt} = k_1 c_A - k_2 c_B c_X + k_3 c_X^2 c_Y - k_4 c_X \tag{5.7}$$

$$\frac{dc_Y}{dt} = k_2 c_B c_X - k_3 c_X^2 c_Y \tag{5.8}$$

where $k_1, k_2, k_3, k_4$ are kinetic rate constants and $c_A$, $c_B$ are concentrations of species $A$ and $B$ that are fixed. Figure 5.8 shows phase trajectories of the Brusselator model for three different initial conditions. While starting from different points, all trajectories converge to a unique closed trajectory—the *limit cycle*.

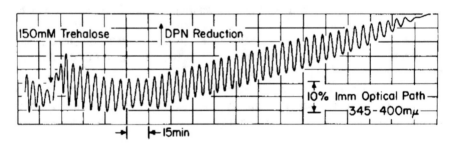

**Fig. 5.6**  Optical recording of periodic oscillations in the cell-free extract of yeast after an addition of trehalose. Reproduced with permission from [10]

**Fig. 5.7**  Phase trajectories of the Lotka–Volterra model. Three trajectories corresponding to different initial conditions are shown. The parameters are $k = k_1 = k_2 = 1$. *Dots* indicate the initial conditions and the *arrow* shows the direction of motion

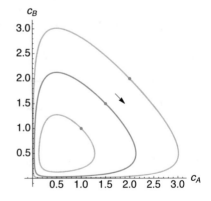

**Fig. 5.8** Phase trajectories of the Brusselator model for three different initial conditions. The parameters are $k_1 = k_2 = k_3 = 1$, $c_A = 1$ and $c_B = 4$. *Dots* indicate the initial conditions and the *arrow* shows the direction of motion

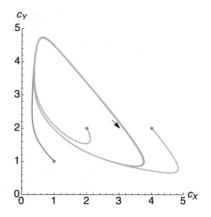

A stable limit cycle represents an attractive periodic orbit, deviations from which vanish with time. Dynamical systems with stable limit cycles have remarkable properties: Independent from initial conditions, they evolve to the same final oscillatory state. The amplitude, the shape and the period of asymptotically established oscillations are not sensitive to initial conditions and are determined only by the dynamical system itself. Thus, systems with stable limit cycles are capable of sustained self-oscillations.

A general theory of limit cycles in dynamical systems was proposed in the nineteenth century by the French mathematician Henry Poincaré. Because oscillations play a fundamental role in electronics, providing the basis for telecommunication devices and radars used in air defence, the theory of oscillatory phenomena rapidly developed in the first half of the twentieth century, before and during the Second World War. Turing was well familiar with electronic oscillators and he also understood that his stationary self-organized structures represented analogs of self-oscillations observed in them. The difference was that, in his structures, the oscillations were taking place in space, not in time.

As shown in the previous chapter, Turing structures arise through an instability - the Turing bifurcation—of a stationary uniform state. Self-oscillations can also develop through a special bifurcation of a steady state. Traditionally, it is known as the *Hopf bifurcation,* although the name is not completely justified.

In the preface to their book *The Hopf Bifurcation and its Applications* [11], J. E. Marsden and M. McCracken have written:

"Historically, the subject had its origins in the works of Poincaré [12] around 1892 and was extensively discussed by Andronov and Witt [13] and their co-workers starting around 1930. Hopf's basic paper [14] appeared in 1942. Although the term "Poincaré-Andronov-Hopf biifurcation" is more accurate (sometimes Friedrichs is also included), the name "Hopf bifurcation" seems more common, so we have used it. Hopf's crucial contribution was the extension from two dimensions to higher dimensions".

As in the previous chapter, we can consider a chemical system with two components having concentrations $u$ and $v$. It will be assumed that these concentrations are uniform, i.e. independent of space, which can be experimentally realized, for example, by stirring of a solution. The effects of spatial nonuniformity and of diffusion will be taken into account in the next chapter. Hence, the kinetic equations are

$$\frac{du}{dt} = f(u, v), \tag{5.9}$$

$$\frac{dv}{dt} = g(u, v). \tag{5.10}$$

Suppose further that there is a stationary state with concentrations $u_0$ and $v_0$, such that $f(u_0, v_0) = 0$ and $g(u_0, v_0) = 0$. Introducing small deviations $\delta u(x, t) = u(x, t) - u_0$ and $\delta v(x, t) = v(x, t) - v_0$ and linearizing the equations, we obtain

$$\frac{d\delta u}{dt} = a\delta u + b\delta v, \tag{5.11}$$

$$\frac{d\delta v}{dt} = c\delta u + d\delta v, \tag{5.12}$$

similar to Eqs. (4.4) and (4.5). The solutions have the form $\delta u(t) \sim \exp(\lambda t)$, $\delta v(t) \sim \exp(\lambda t)$ where $\lambda = \gamma \pm i\omega$ with $\gamma = (1/2)(a + d)$ and $\omega = (1/2)\sqrt{|(a - d)^2 + bc|}$ provided that $(a - d)^2 + bc < 0$.

The oscillations can only take place if either $b < 0$ and $c > 0$ or $b > 0$ and $c < 0$. Suppose that $a > 0$ and thus autocatalysis is characteristic for the species $u$ which plays the role of an activator. Then the first of the above conditions (i.e., $b < 0, c > 0$) implies that the second species $v$ is effectively an inhibitor that suppresses the autocatalysis of $u$ and is, however, produced by the species $u$. If the second condition (i.e., $b > 0, c < 0$) is instead satisfied, the second species is a "fuel" for the autocatalysis of $u$; this fuel is consumed during the reproduction of the first species. The oscillations are damped if $a + d < 0$ and grow with time if $a + d > 0$. The growth of perturbations means the instability of the stationary state. This instability occurs at a critical point defined by the equation $a + d = 0$. Note that the growth condition $a + d > 0$ implies that either $a$ or $d$, or both these coefficients, are positive. Thus, effective autocatalysis should take place at least for one of the species in order that the instability occurs.

Such instability of the stationary state represents the Hopf bifurcation. It is possible in essentially the same activator-inhibitor systems as the Turing bifurcation, although in a different parameter region.

What will happen above the instability threshold? In the linear system (5.11) and (5.12), the amplitude of oscillations grows indefinitely with time. However, this system of equations holds only for small deviations from the stationary state. If they grow, nonlinear effects described by the original kinetic Eqs. (5.9) and (5.10) should come into play. If nonlinear effects are slowing down the growth of perturbations, small-amplitude persistent oscillations arise.

Such oscillations can be described by a general model where only the first non-linearities are taken into account. To formulate this model, a complex oscillation amplitude $\eta(t)$ is introduced, such that the deviations from the stationary state can be expressed as $\delta u(t) = A\eta(t) + c.c.$ and $\delta v(t) = B\eta(t) + c.c.$ where the notation $c.c.$ means taking the complex conjugate of the first terms. In the weakly nonlinear regime, complex oscillation amplitudes approximately satisfy the equation

$$\frac{d\eta}{dt} = \gamma\eta + i\omega\eta - (\alpha + i\beta)\,|\eta|^2\,\eta \tag{5.13}$$

where coefficients $\alpha$ and $\beta$ are determined from the full kinetic Eqs. (5.9) and (5.10).

The model (5.13) plays an important role in the theory of oscillations. Essentially, this is the so-called *normal form* for the Hopf bifurcation (the theory of normal forms for bifurcations of stationary states of differential equations was constructed by Poincaré). In physical literature, it is also sometimes called the "Stuart–Landau equation".

Its reformulation leads to

$$\frac{d\eta}{dt} = \left(\gamma - \alpha\,|\eta|^2\right)\eta + i\left(\omega - \beta\,|\eta|^2\right)\eta. \tag{5.14}$$

Thus, the term with the coefficient $\alpha$ in Eq. (5.13) corresponds to a nonlinear change of the growth rate $\gamma$, whereas the term with $\beta$ yields a nonlinear modification of the oscillation frequency $\omega$.

Near the critical point, the coefficient $\gamma$ is small and therefore even weak nonlinearities are enough to affect it and to change its sign. If the magnitude $|\eta|$ of the oscillation amplitude is equal to

$$\rho_0 = \sqrt{\frac{\gamma}{\alpha}} \tag{5.15}$$

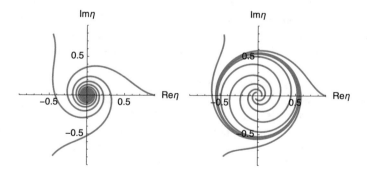

**Fig. 5.9** The Hopf bifurcation. Phase trajectories of Eq. (5.13) at $\gamma = -0.1$ (*left panel*) and $\gamma = 0.5$ (*right panel*). Other parameters are $\alpha = \beta = \omega = 1$

the oscillations neither grow nor decay; hence persistent oscillations with this amplitude will be observed. The frequency of such oscillations is $\omega_0 = \omega - \beta\gamma/\alpha$. Furthermore, oscillations with this amplitude and frequency are stable. Indeed, if $|\eta|$ is smaller than $\rho_0$, the effective growth rate $\gamma_{eff} = \gamma - \alpha|\eta|^2$ is positive and the oscillations grow. On the other hand, if $|\eta|$ is larger than $\rho_0$, $\gamma_{eff}$ is negative and the amplitude of oscillations shrinks.

Hence, when $\gamma > 0$ the system has a attractive limit cycle representing, in the variables $\mathrm{Re}\,\eta(t)$ and $\mathrm{Im}\eta(t)$, a circle of radius $\rho_0$. For $\gamma < 0$, the limit cycle is absent and the stationary point is stable. The Hopf bifurcation describes the birth of a limit cycle from an attractive stationary state. As illustrated in Fig. 5.9, this state becomes unstable and gives rise to a small limit cycle that surrounds it.

The characteristic feature of the Hopf bifurcation is that the amplitude of oscillations is small at the critical point and the oscillations are approximately harmonical. The amplitude of oscillations increases as $\gamma^{1/2}$ when the bifurcation parameter $\gamma$ is increased. The period of oscillations remains approximately constant. The deviations from the limit cycle are damped and the respective relaxation rate is proportional to $\gamma$, going to zero at the bifurcation point.

Above we have assumed that the coefficient $\alpha$ is positive. As we have seen, weak oscillations develop in this *supercritical* case when the threshold $\gamma = 0$ is crossed. If, however, $\alpha < 0$, the situation is different. Then, $\gamma_{eff} = \gamma - \alpha|\eta|^2$ grows under an increase of the oscillation amplitude and thus the instability is accelerated. Hence, the "catastrophal" instability takes place. In this case, Eq. (5.13) describes only the first stage of the instability development and full kinetic Eqs. (5.9) and (5.10) have to be employed to consider the final oscillatory regime. Note that in this latter case the stationary state is unstable with respect to perturbations with a sufficiently strong amplitude even before the formal instability threshold $\gamma = 0$ has been reached. Therefore, the bifurcation is *subcritical* then.

The Hopf bifurcation is not the only instability that can give rise to a limit cycle (or describe the disappearance of it). For example, the limit cycle can also disappear through the so called *saddle-node bifurcation on the invariant curve*. In this case, the motion along a limit cycle becomes progressively slower near some position on it as the control parameter is changed. Eventually, at the bifurcation point, the motion vanishes at this position and thus a stationary state becomes formed. The characteristic property of this bifurcation is that the oscillation period diverges as the critical point is reached.

Dynamical systems with attractive limit cycles have special properties. Although they may have many degrees of freedom, their dynamics is essentially one-dimensional and proceeds on a "rail track" representing the limit cycle. The coordinate along such a track is the running oscillation phase $\phi$. It can be defined through the relative time needed to reach a given point on the cycle. Thus, by definition, the phase velocity is constant along the cycle and the equation

$$\frac{d\phi}{dt} = \omega_0 \tag{5.16}$$

holds where $\omega_0 = 2\pi/T_0$ is the natural frequency of oscillations.

In terms of the phase, responses of oscillators to external perturbations can readily be rationalized. Generally, a perturbation will have two effects. First, it will move the system a little away from its limit-cycle track. The deviation will be however small if the perturbation is sufficiently weak (so that it does not kick the system completely off its rail track). Moreover, the perturbation will also modify the local phase velocity, so that instead of Eq. (5.16) we will have

$$\frac{d\phi}{dt} = \omega_0 + g(\phi)p(t). \qquad (5.17)$$

Here, $g(\phi)$ is the *sensitivity function* that describes the response to a perturbation at a given phase along the cycle; this function depends on the dynamical system and on the kind of the perturbation applied. The parameter $p$ specifies the magnitude of the perturbation; it can depend on time if the perturbation is not constant or is applied only for a while.

Suppose that a short perturbation of small magnitude $\varepsilon$ has been applied at some time $t_0$, so that $p(t) = \varepsilon\delta(t - t_0)$. According to Eq. (5.17), this would lead to a phase shift $\Delta\phi = \varepsilon g(\phi(t_0))$ in the motion along the cycle. Although this shift is small, it does not disappear as time goes on. Therefore, phase changes induced by perturbations will become *accumulated* with time.

Interactions between oscillators can be considered in a general way when the phase description is used. This has been done by Arthur Winfree [15] in 1966 while he was working on his PhD thesis in theoretical biology at Princeton university. Stressing that rhythms are ubiquitous in biological systems, he has asked: "What special phenomena can we expect to arise from the rhythmical *interaction* of whole populations of periodic processes? How could we *recognize* a multi-oscillator community if encountered, and how probe its integrative mechanisms ?" As he remarked, this would be a community "knit together by periodic interactions, as in laser".

The action of an oscillator onto another oscillator can be viewed as a perturbation affecting the phase motion of the first oscillator. Therefore, it can be described by Eq. (5.17). The perturbation made by oscillator 2 on oscillator 1 can be expressed as $p(t) = \varepsilon f(\phi_2(t))$ where the parameter $\varepsilon$ specifies the strength of the interaction and the *influence function* $f(\phi_2)$ characterizes the variation in the generated force depending on the phase $\phi_2$ of the oscillator 2. Thus, the system of two interacting identical oscillators will be described by equations

$$\frac{d\phi_1}{dt} = \omega_0 + \varepsilon g(\phi_1)f(\phi_2), \quad \frac{d\phi_2}{dt} = \omega_0 + \varepsilon g(\phi_2)f(\phi_1). \qquad (5.18)$$

The extension of this description to a population of $N$ interacting oscillators with different natural frequencies $\omega_i$ yields the *Winfree model*:

$$\frac{d\phi_i}{dt} = \omega_i + \varepsilon \sum_{j=1, j\neq i}^{N} g(\phi_i)f\left(\phi_j\right) \qquad (5.19)$$

The principal effect described by Eq. (5.19) is *synchronization* of oscillator dynamics. It was investigated by Winfree [15] using the Princeton IBM 7094 digital computer, one of the most powerful at that time.

Figure 5.10 shows four consequent snapshots from a simulation demonstrating the development of a synchronous state from the initial random distribution of phases. The population consisted of 100 oscillators with different frequencies. In the figure, instantaneous phases $\varphi_i = \phi_i/2\pi$ of oscillators are plotted against their natural periods $T_i = 2\pi/\omega_i$.

At time zero in Fig. 5.10a, oscillators are all running at their own rates because the phases are initially random and there is no net rhythm in the community. After 1000 time units (Fig. 5.10b), condensation in the cloud of phases is already observed. At the time of 2000 (Fig. 5.10c), the emergence of a coherent group is clear, although some oscillators on the flanks are not yet entrained.

In the last snapshot (Fig. 5.10d), almost perfect mutual synchrony is established. In this synchronized states, all oscillators have the same constant phase velocity (i.e., the collective oscillation frequency), although their oscillation phases are different and depend on the natural individual frequency in each of them. Winfree has also constructed an approximate analytical theory that gave conditions for complete synchronization.

The Winfree model is tailored to the situations where different oscillation phases correspond to largely different physical states, so that the sensitivity and the influence functions of an oscillator both depend on its own phase state. For harmonic oscillations, this condition does not seem natural. The limit cycle is then circular and has no dependence on the phase. Hence, the sensitivity and the influence functions should probably not depend on the absolute phase too. However, interactions between two such oscillators may still well depend on the difference of their phases.

Yoshiki Kuramoto, motivated by the work of Winfree, has proposed [16] in 1975 a different equation for interactions between oscillators,

$$\frac{d\phi_i}{dt} = \omega_i + \varepsilon \sum_{j=1}^{N} \Gamma\left(\phi_i - \phi_j\right) \tag{5.20}$$

where $\Gamma(\phi)$ is the interaction function; often it is chosen as $\Gamma(\phi) = \sin\phi$. The equation is known as the *Kuramoto model*.

What is the relationship between the Kuramoto and Winfree models? Suppose that $\Gamma(\phi) = \sin\phi$. Then, by using the identity $\sin(\phi_i - \phi_j) = \sin\phi_i \cos\phi_j - \sin\phi_j \cos\phi_i$, we can write Eq. (5.20) in the form

$$\frac{d\phi_i}{dt} = \omega_i + \varepsilon \sum_{j=1}^{N} \sin\phi_i \cos\phi_j - \varepsilon \sum_{j=1}^{N} \cos\phi_i \sin\phi_j \tag{5.21}$$

Now, it can be seen that, in the Kuramoto model in this case, there are two parallel communication channels between any pair of oscillators $i$ and $j$. In the first of them, the sensitivity function is $\sin\phi$ and the influence function is $\cos\phi$, whereas in the

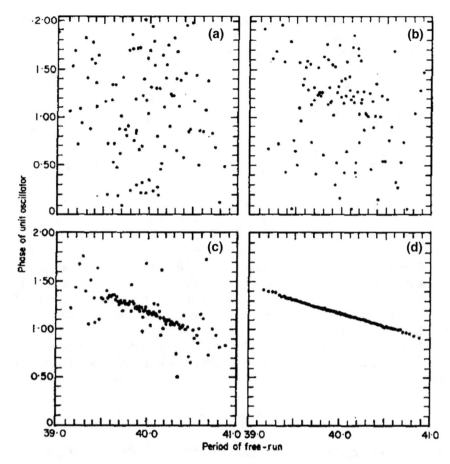

**Fig. 5.10** Synchronization in a population of 100 interacting phase oscillators. Subsequent snapshots at times **a** 0, **b** 1000, **c** 2000 and **d** 3000 are shown. The phases are plotted against the natural oscillation periods of oscillators, i.e. $T_i = 2\pi/\omega_i$. Reproduced with permission from [15]

second channel we have the sensitivity function $-\cos\phi$ and the influence function $\sin\phi$.

Similar to Eq. (5.19), the Kuramoto model can show synchronization, clustering and desynchronization depending on the strength of interactions and the function $\Gamma(\phi)$. However, it is simpler than the Winfree model and more amenable for theoretical analysis.

If oscillators are identical, they synchonize according to Eq. (5.20) at any interaction strength $\varepsilon$ if the condition $d\Gamma/d\phi < 0$ is satisfied at $\phi = 0$; if the opposite condition holds, the synchronous state is unstable. The synchronization can also be achieved if oscillators are heterogeneous and have different natural frequencies, provided that the interactions are strong enough. At lower interaction strengths, partial synchronization is possible where only a subset of oscillators forms a coherently oscillating group, whereas other oscilators remain non-entrained.

In his classical book *Chemical Oscillations, Waves and Turbulence* [17] published in 1984, Kuramoto has developed a statistical theory of the synchronization transition in heterogeneous oscillator publications as a critical phenomenon. When the interaction strength $\varepsilon$ is gradually increased, first a small synchronous group of oscillators emerges above the critical point. Upon further increase of $\varepsilon$, the size of this group grows and, eventually, it encompasses all oscillators, so that complete synchronization is achieved.

To quantitatively characterize synchronization, Kuramoto introduced the order parameter

$$Z = \frac{1}{N} \sum_{i=1}^{N} e^{i\phi_i} \qquad (5.22)$$

If phases $\phi_i$ of all oscillators are independent and random, this order parameter is (almost) zero for large N. On the other hand, if all phases are identical, i.e. $\phi_i = \Phi$, we have $Z = e^{i\Phi}$ and therefore $|Z| = 1$. Thus, the absolute value $R = |Z|$ of the synchronization parameter provides a measure for phase synchronization in an oscillator population.

If natural frequencies of oscillators have the Lorenz distribution with the center at $\omega = \omega_0$ and width $\delta$, i.e. the probability to have an oscillator with natural frequency $\omega$ is

$$p(\omega) = \frac{\delta/\pi}{(\omega - \omega_0)^2 + \delta^2}, \qquad (5.23)$$

and $\Gamma(\phi) = \sin\phi$, the theory of Kuramoto predicts that the critical point is reached at the interaction intensity $\varepsilon_c = 2\delta$. Below it, synchronization is absent ($R = 0$) and above it the order parameter depends on $\varepsilon$ as

$$R = \sqrt{1 - \frac{\epsilon_c}{\varepsilon}} \qquad (5.24)$$

Thus, near the critical point, $R$ is proportional to $(\varepsilon - \varepsilon_c)^{1/2}$. This proportionality law is general and holds also for other frequency distributions near the respective critical points.

Above the synchronization transition, all oscillators with the natural frequencies in the interval from $\omega_0 - \varepsilon R$ to $\omega_0 + \varepsilon R$ build up a coherent group. Instead of rotating at individual natural frequencies $\omega_i$, they move at the collective frequency $\Omega = \omega_0$. Note that only for symmetric frequency distributions the collective frequency $\Omega$ coincides with the frequency of the peak.

To apply such theoretical results to experimental systems, one needs first to decide how the oscillation phase can be determined. The above definition (5.16) was mathematical and referred to the motion along a periodic attractive trajectory, i.e. along the limit cycle, in the full space of all variables of a dynamical system. In the experiments, it is however often that only one property, such as, e.g., a certain reaction rate $s(t)$, can be continuously monitored. The question is how to reconstruct the limit cycle and find oscillation phases in this case.

Generally, the idea is that, as two independent variables, the rates $s(t)$ at a given moment $t$ and $q(t) = s(t - \Delta)$ at a delayed time moment $t - \Delta$ can be chosen. When the signal is plotted depending on time on the plane $(s, q)$, motion along a limit cycle may already be seen, so that the phase can be defined.

In the applications, a somewhat different method known as the *analytic signal approach* [18] (see also [19]) is preferred. Instead of the delayed signal, its Hilbert transform

$$\widetilde{s}(t) = \frac{1}{\pi} \int_{-\infty}^{\infty} \frac{s(t')}{t - t'} dt' \tag{5.25}$$

is used (this can be done by taking the Fourier transform of $s(t)$, shifting each complex Fourier coefficient by a phase of $\pi/2$ and performing the reverse Fourier transform). Using $s(t)$ and its Hilbert transform $\widetilde{s}(t)$, a complex variable $\zeta(t) = s(t) + i\widetilde{s}(t)$, known as the analytic signal, is produced.

While the synchronization theory of Kuramoto became well known and broadly discussed, its verification for chemical systems came only in 2002 in the experiments [20] by Istvan Kiss, Yumei Zhai and John L. Hudson in the University of Virginia. Previously, it had been found [21] that electrochemical dissolution of nickel in sulphuric acid yielded rate oscillations. The designed experimental system represented an array of 64 Ni electrodes immersed in sulphuric acid; they were integrated into an electrical circuit in such a way that uniform coupling between them was realized.The strength $K$ of coupling, determining the strength of interactions between the oscillators, could be electrically controlled.

The current, which is proportional to the rate of metal dissolution, was measured on each electrode $j$ and almost harmonic oscillations were observed (Fig.5.11a). The natural frequencies of individual oscillators in absence of coupling were a little different. To display oscillations, the current $i(t)$ and its Hilbert transform were employed. Figure 5.11b shows a snapshot of the states of all 64 oscillators in the analytic signal plane, in a situation when coupling between them was switched off. As can be seen, the oscillators stay on a limit cycle of an elliptic shape and are almost uniformly distributed along it.

For all oscillators $j$, their effective oscillation frequencies $\omega_j(K)$ at different interaction strength $K$ were determined. Figure 5.12a shows such frequencies as a function of natural frequencies of the oscillators $\omega_j(K = 0)$ above the critical point. The central group of oscillators has the same frequency, whereas the oscillators on the flanks are non-entrained.

The synchronization order parameter $r$ was determined in these experiments as the vector sum of the phase points in the analytical signal plane, similar to that defined by Eq. (5.22). The dependence of the time-averaged order parameter on the coupling strength is shown in Fig. 5.12b. It starts to rapidly increase above the critical point

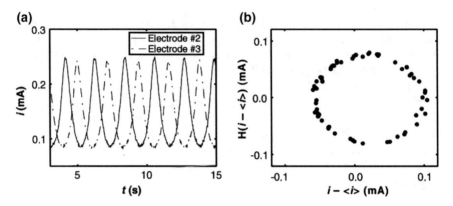

**Fig. 5.11** Oscillations in electrochemical dissolution of Ni in sulphuric acid. **a** Currents, proportional to rate of metal dissolution, for two selected electrodes. **b** Instantaneous states of all 64 oscillators in the plane of the variables representing the current and its Hilbert transform. Reproduced with permission from [20]

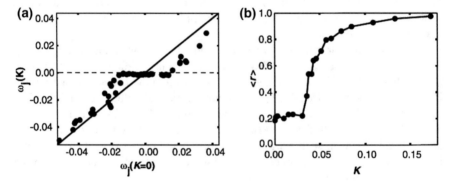

**Fig. 5.12** Synchronization of electrochemical oscillators. **a** Effective oscillation frequencies of oscillators versus their natural frequencies at the coupling strength $K = 0.034$, above the critical coupling strength of $K_c = 0.03$. **b** The mean syncronization order parameter as a function of the coupling strength $K$. Reproduced with permission from [20]

$K_c = 0.03$ and this increase can be approximated by the theoretical dependence (5.24), replacing $\varepsilon$ by $K$. Below the critical point, it does not however vanish as predicted by the theory. This is because the theory was constructed in the limit of an infinitely large number of oscillators, whereas only 64 oscillators were present in the experiments. The authors showed that the remaining order parameter scaled as $N^{-1/2}$ when the number $N$ of oscillators was increased.

An important mathematical discovery of the twentieth century was that even simple nonlinear dynamical systems can show intrinsically chaotic dynamics. In 1963, Edward Lorenz derived [22] a model of three coupled nonlinear differential equations while studying the problem of hydrodynamic convection. Simulating this dynamical system on an analog computer, he has found that, surprisingly, the solutions were sustainably irregular, without any tendency to converge to periodic or quasi-periodic oscillations, or to die out, despite the fact that no random external perturbations were applied.

The distinguishing property of chaotic oscillations is their extremely high sensitivity to perturbations: separation between two initially close chaotic trajectories grows exponentially with time. Such sensitivity can be quantified by introducing

$$\lambda = \lim_{t \to \infty} \frac{1}{t} \ln(d(t)/d(0)) \tag{5.26}$$

where $d(t)$ is the distance at time $t$ between two system's trajectories that were initially separated by the distance $d(0)$. This quantity is known as the *Lyapunov exponent*. Its positive sign signals the onset of chaos.

The first chemical model exhibiting chaotic oscillations was constructed by Otto Rössler in 1976 [23]. The model had three reactants and kinetic equations for their concentrations $u$, $v$ and $w$ were

$$\frac{du}{dt} = k_1 + k_2 u - \frac{(k_3 v + k_4 c) u}{u + K} \tag{5.27}$$

$$\frac{dv}{dt} = k_5 u - k_6 v \tag{5.28}$$

$$\mu \frac{dw}{dt} = k_7 u + k_8 w - k_9 w^2 - k_{10} \frac{w}{w + K'} \tag{5.29}$$

These equations corresponded to a hypothetical reaction scheme and were derived by adiabatic elimination of fast intermediate products. By numerical integration, Rössler showed that, for some parameter choices, persistent irregular oscillations were observed. In the limit $\mu \to 0$, he could also analytically prove the existence of chaotic oscillations.

Somewhat later, he has also proposed [24] a different mathematical system, representing a simplification of the original Lorenz equations and known as the *Rössler model*. This system does not correspond to any set of chemical reactions and its equations are

$$\frac{dx}{dt} = -y - z \tag{5.30}$$

$$\frac{dy}{dt} = x + ay \tag{5.31}$$

$$\frac{dz}{dt} = b + z(x - c) \tag{5.32}$$

A remarkable property of this model is that it includes only one nonlinear term in the equation for the variable $z$.

**Fig. 5.13** A phase trajectory
of the Rössler model for the
parameters $a = 0.2$, $b = 0.2$
and $c = 5.7$

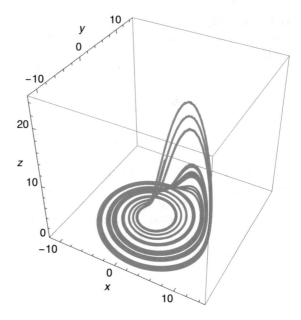

Figure 5.13 shows an example of the phase trajectory obtained by numerical integration of Eqs. (5.30)–(5.32) in the chaotic regime. There is an unstable stationary state in the plane $(x, y)$ and the trajectory spirals out of it. When the deviations from this state become strong, the variable $z$ rises and the trajectory moves up from the plane $(x, y)$, but is then injected back towards the unstable stationary point, so that the next cycle can begin. In Fig. 5.14, the chaotic oscillations in the variable $z$ are displayed. A similar behavior can be also found in the chemical model (5.27), (5.29).

The transition to chaotic oscillations in the Rössler model takes place through a sequence of *period-doubling bifurcations* (Fig. 5.15). First, a simple limit cycle is found (Fig. 5.15, $c = 2.6$). As the bifurcation parameter $c$ is increased, it becomes transformed to a twice-coiled limit cycle with the double period (Fig. 5.15, $c = 3.5$). Further doublings lead to the cycles of period four (Fig. 5.15, $c = 4.1$) and more. Eventually, the period tends to infinity and this corresponds to the emergence of chaotic dynamics (Fig. 5.15, $c = 4.23$). The period-doubling scenario of transition to chaos is characteristic for many models and it is also found in the experiments (an example will be shown in Chap. 8). Other scenarios are however also possible and have been observed.

First experimental observations of chaotic oscillations were reported in 1977 by Lars Olsen and Hans Degn [25] in the experiments on oxidation of NADH by $O_2$ catalyzed by horseradish peroxidase. The reaction was taking place in a stirred flow reactor into which NADH was pumped at a constant rate; the reactor was open for $O_2$. This work was motivated by the theoretical study of Rössler and the same mathematical tools [24], based on the analysis of the recurrence map, were used to demonstrate the chaotic character of observed oscillations. In the same year, chaotic

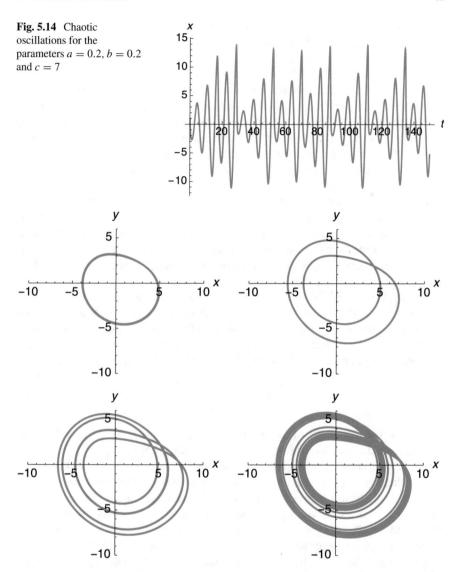

**Fig. 5.14** Chaotic oscillations for the parameters $a = 0.2$, $b = 0.2$ and $c = 7$

**Fig. 5.15** Transition to chaos through period-doubling bifurcations in the Rössler model: $c = 2.6$ (*top left*), $c = 3.5$ (*top right*), $c = 4.1$ (*bottom left*), $c = 4.23$ (*bottom right*). Other parameters are $a = 0.2$ and $b = 0.2$. Projections of three-dimensional trajectories on the plane $(x, y)$ are shown

oscillations were found in the Belousov-Zhabotinsky reaction [26]. We shall return to the discussion of chaotic oscillation regimes in Chap. 7 where the Belousov-Zhabotinsky reaction is introduced.

It should be stressed that the irregularity of experimentally observed oscillations does not itself imply that they are intrinsically chaotic. Indeed, non-controlled variations of external parameters may be present in an experiment. Moreover, the oscillations can be affected by noise (particularly on microscales where fluctuations are relatively strong). There are however statistical methods that allow to distinguish true chaotic oscillations from randomly fluctuating signals (see Chap. 7).

The theory of dynamical chaos is nowadays a well established mathematical field. Its detailed presentation would have lead us too far into the realm of mathematics and we do not undertake this in the present book focused on the basic effects. Before concluding this chapter, we want however to note that synchronization, similar to that we have described above, is also possible in populations of coupled chaotic oscillators. In the experiments with electrochemical dissolution of Ni in sulphuric acid [20], chaotic oscillations were taking place for some parameter choices and their synchronization could be clearly seen.

Finally, the response of periodic limit-cycle oscillations to external periodic forcing can also be discussed. For damped oscillations in linear systems, resonance phenomena are found. If the natural frequency of such oscillations is $\omega_0$, the stimulation frequency is $\omega$, and the damping rate constant is $\gamma$, the amplitude $A$ of forced oscillations depends resonantly on the difference between $\omega$ and $\omega_0$, that is

$$A \propto \frac{f\gamma}{(\omega - \omega_0)^2 + \gamma^2} \tag{5.33}$$

where $f$ is the magnitude of the applied force. Thus, it is sharply increased if the forcing frequency is close to that of the natural oscillations.

For limit-cycle oscillations, the amplitude and the shape are fixed and they cannot be significantly changed by external forcing (unless it is so strong that it destructively interferes with the self-organization process). Therefore, resonances in the amplitude of such oscillations cannot occur. Instead, the effects of *entrainment* by external forcing take place. The periodic limit-cycle motion becomes accelerated or slowed so that its actual frequency become equal to the forcing frequency $\omega_0$. This is however only possible if the forcing frequency is not much different from the natural frequency of limit cycle oscillations $\omega_0$. Moreover, the forcing should be also strong enough. Thus, for each force $f$, the entrainment window

$$\omega_0 - \nu_-(f) \leq \omega \leq \omega_0 + \nu_+(f) \tag{5.34}$$

can be defined. This window is characterized by the functions $\nu_-(f)$ and $\nu_+(f)$ of the force strength $f$. The entrainment window shrinks as $f$ is decreased and vanishes when $f = 0$. In the literature, it is known as the *Arnold tongue* (Vladimir Arnold was a Russian mathematician who has worked much on the theory of nonlinear dynamics).

Generally, the entrainment can occur at the forcing frequencies $(k/l)\omega_0$ with any integers $k, l = 1, 2, 3....$ If $k > l$ the entrainment is superharmonic; whereas if $k < l$ it is subharmonic The synchronization windows are however more narrow in such $k{:}l$ entrainment regimes than in the case of the harmonic entrainment with $k = l = 1$.

# References

1. W. Ostwald, Z. Phys. Chem. **25**, 33 (1899)
2. A.J. Lotka, J. Phys. Chem. **14**, 271 (1910)
3. A.J. Lotka, Proc. Natl. Acad. Sci. USA **6**, 410 (1920)
4. V. Volterra, *Leçons sur la Theorie Mathématique de la Lutte pour la Vie* (Gautier, Paris, 1931)
5. E.P. Odun, *Fundamentals of Ecology* (Saunders, Philadelphia, 1953)
6. W.C. Bray, J. Am. Chem. Soc. **43**, 1262 (1921)
7. A.M. Turing, Philos. Trans. R. Soc. Lond. B **237**, 37 (1952)
8. L.N. Duysens, J. Amesz, Biochim. Biophys. Acta **24**, 19 (1957)
9. A. Ghosh, B. Chance, Biochem. Biophys. Res. Commun. **16**, 174 (1964)
10. B. Hess, K. Brand, K. Pye, Biochem. Biophys. Res. Commun. **23**, 102 (1966)
11. J.E. Marsden, M. McCracken, *The Hopf Bifurcation and Its Applications* (Springer, New York, 1976)
12. H. Poincaré, *Les Méthodes Nouvelles de la Mécanique Céleste*, vol. 1 (Gauthier-Villars, Paris, 1892)
13. A.A. Andronov, A. Witt, Acad. Sci. Paris **190**, 256 (1930)
14. E. Hopf, Abzweigung einer periodischen Losung von einer stationarer Losung eines Differentialsystem. Ber. Math.-Phys. Sachsische Akademie der Wissenschaften Leipzig **94**, 1 (1942)
15. A. Winfree, J. Theor. Biol. **16**, 15 (1967)
16. Y. Kuramoto, in *International Symposium on Mathematical Problems in Theoretical Physics*, ed. by H. Araki, in Springer Lecture Notes in Physics, vol. 39, (Springer, New York, 1975), p. 420
17. Y. Kuramoto, *Chemical Oscillations, Waves and Turbulence* (Springer, Berlin, 1984)
18. P. Panter, *Modulation, Noise and Spectral Analysis* (McGraw-Hill, New York, 1965)
19. M. Rosenblum, J. Kurths, in *Nonlinear Analysis of Physiological Data*, ed. by H. Kantz, J. Kurths, G. Mayer-Kress (Springer, Berlin, 1998), p. 91
20. I. Kiss, Y. Zhai, J.L. Hudson, Science **296**, 1676 (2002)
21. O. Lev, A. Wolfberg, M. Sheintuch, L.M. Pismen, Chem. Eng. Sci. **43**, 1339 (1988)
22. E.N. Lorenz, J. Atm. Sci. **20**, 130 (1963)
23. O.E. Rössler, Z. Naturforschung **31a**, 259 (1976)
24. O.E. Rössler, Phys. Lett. A **57**, 397 (1976)
25. L.F. Olsen, H. Degn, Nature **267**, 177 (1977)
26. R.A. Schmitz, K.R. Graziani, J.L. Hudson, J. Chem. Phys. **67**, 3040 (1977)

# Chapter 6
# Propagating Waves

At the beginning of the 20th century, it was known that electric pulses could propagate without damping along nerves over the distances of meters at the speed of about 50 m/s. While this phenomenon was of fundamental importance, its mechanism remained at that time unclear. Wilhelm Ostwald suggested that initiation and propagation of crystallization waves in a supercooled melt might be an analog of the nerve propagation phenomena—but the nerve was obviously not a solid, and therefore other effects had to be involved. In a talk [1] held in 1906 at the meeting of Bunsengesellschaft für Physikalische Chemie in Dresden, Robert Luther, a professor from Leipzig and the former associate of Ostwald, wanted to demonstrate that traveling waves exist in liquid homogeneous systems and thus another mechanism for nerve propagation was possible.

As Luther remarked, "in studying the actual cause of the propagation of crystallizations, one soon arrives at the conclusion that the processes that propagate in a homogeneous medium have to be autocatalytic". Therefore, the reaction of autocatalytic decomposition of the salts of alkyl sulphonic acids was chosen. The experiments were performed in tubes filled with the solution of methyl or ethyl sulphate; to prevent convection the solution was gelled. Explaining the expected effects, R. Luther noted:

> The neutral alkyl sulphates are very stable in aqueous solutions but slowly give off sulphuric acid upon acidification. If I introduce some acid at one end of the tube then, under the catalytic influence of $H^+$, hydrolysis ensues and new $H^+$ ions are formed. These ions diffuse to the right and cause the formation of more acid. In this way, decomposition slowly moves through the tube.

Thus, a propagating $H^+$ would be formed and its motion can be observed by using a color indicator. Several systems with autocatalytic reactions were considered and, for a demonstration during the talk, the reaction between permanganate $KMnO_4$ and oxalic acid had been used. A front propagating through the tube at the velocity of several centimeters per hour was observed.

In the same talk, Luther also showed a formula for the propagation velocity of a front,

© Springer International Publishing AG 2017
A.S. Mikhailov and G. Ertl, *Chemical Complexity*, The Frontiers Collection,
DOI 10.1007/978-3-319-57377-9_6

$$V = a\sqrt{kcD} \qquad (6.1)$$

where $k$ was the reaction rate constant, $c$ was the (substrate) concentration, $D$ was the diffusion constant, and $a$ was a numerical coefficient. This result was said to be "a simple consequence of the corresponding differential equation", but the equation had not been however presented and the derivation of this formula was not given. It might be that Luther had arrived at the expression (6.1) by using dimensionality arguments.

It was later shown by Alan Hodgkin and Andrew Huxley that propagation of electric pulses along nerves relies on a different mechanism, involving ionic currents passing through biological membranes. Nonetheless, the work by Luther had shown that traveling self-supporting fronts are possible in homogeneous reactions in aqueous solutions. In 1987, the English translation of his article was published together with the accompanying comments [2].

Already Alfred Lotka noted [3] that autocatalysis is analogous to the process of biological reproduction which can be described by the equation proposed in 1838 by the Belgian mathematician Pierre-François Verhulst [4] and also known as the equation of *logistic growth*,

$$\frac{dN}{dt} = rN\left(1 - \frac{N}{K}\right), \qquad (6.2)$$

where $N$ is the number of the reproducing species, $r$ is the reproduction rate, and $K$ is the population number in the final stationary state. According to this equation, the population first grows exponentially, but then the growth is saturated and a steady state is reached.

In the thirties of the 20th century, mathematical genetics was developed. To describe spreading of advantageos genes, the British statistician and biologist Ronald Fisher proposed [5] in 1937 an equation that extends (6.2) by including a diffusion term,

$$\frac{\partial u}{\partial t} = au - bu^2 + D\frac{\partial^2 u}{\partial x^2} \qquad (6.3)$$

where $u$ is the local population density (of the reproducing gene), $a$ is the reproduction rate, $b$ is a constant that determines the saturation density, and $D$ is the diffusion coefficient. Fisher showed that this equation has solutions in the form of fronts traveling at a constant speed and the propagation velocity is greater or equal to $V = 2\sqrt{aD}$. He argued that this lowest velocity should be the "natural" speed at which the fronts are to be observed, but could not prove this.

Figure 6.1, obtained by numerical integration of Eq. (6.3), shows how a traveling front develops from an initial localized perturbation and begins to propagate at a constant velocity while retaining its shape.

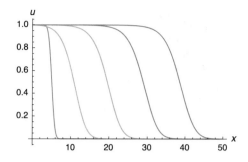

**Fig. 6.1** A traveling front. Numerical integration of Eq. (6.3) with $a = b = D = 1$. Snapshots of the population distribution at consequent moments separated by time intervals $\Delta t = 5$ are displayed

In the same year, Russian mathematicians Andrey Kolmogorov, Ivan Petrovsky and Nikolai Piskunov, also interested in genetics, considered [6] a general equation

$$\frac{\partial u}{\partial t} = f(u) + D\frac{\partial^2 u}{\partial x^2} \tag{6.4}$$

In this equation, $f(0) = 0$ and $f'(0) > 0$, so that the stationary state $u = 0$ is unstable. They have shown that, if the population was initially localized within a region in the medium, it spreads through it as a front traveling at the velocity $V = 2\sqrt{f'(0)D}$. There are also front solutions traveling at a greater velocity, but they correspond to the situations when the medium was initially everywhere populated, perhaps at a low level, by the reproducing species.

Although the Fisher equation (6.2) and the Kolmogorov-Petrovsky-Piskunov (KPP) equation (6.4) were formulated in the context of genetics, they also describe spreading of infections and are thus used in epidemiology. Another application where such equations arise is provided by combustion phenomena.

In the latter case, the concentration $c$ corresponds to the local temperature $\theta$ and the role of diffusion is played by heat conduction. The rate of heat release $q$ is proportional to the rate of the fuel oxidation reaction that has the Arrhenius dependence on temperature, so that $q(\theta) = W\exp(-E_a/k_B\theta)$ where $E_a$ is the activation energy and $W$ is a coefficient proportional to the fuel concentration (assumed to be constant). If a heat amount $q\Delta t$ is released within time $\Delta t$, the temperature will increase by $\Delta\theta = q\Delta t/C$ where $C$ is the heat capacity. Thus, the rate of local temperature change due to combustion is $d\theta/dt = q(\theta)/C$. In addition, heat exchange with the environment and the process of heat conduction within the system will take place. Hence, the evolution equation for temperature distribution will be

$$\frac{\partial \theta}{\partial t} = f(\theta) + \chi\frac{\partial^2 \theta}{\partial x^2} \tag{6.5}$$

with

$$f(\theta) = \frac{W}{C}\exp\left(-\frac{E_a}{k_B\theta}\right) - \gamma(\theta - \theta_0) \tag{6.6}$$

**Fig. 6.2** Typical function
$f(u)$ for bistable chemical
media. The sign of the
difference of the two *shaded
areas* determines the
propagation direction of the
front

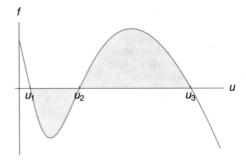

where $\gamma$ is the coefficient characterizing heat exchange with the environment at
temperature $\theta_0$ and $\chi$ is the thermal conductivity of the medium.

A simple example of a chemical system with an autocatalytic reaction was proposed in 1982 by Friedrich Schlögl [7]. He has considered a reaction scheme

$$A + 2X \rightleftharpoons 3X,\ X \rightleftharpoons B \tag{6.7}$$

According to it, the concentration $u$ of species $X$ should satisfy, with inclusion of
diffusion, Eq. (6.4) with the function

$$f(u) = k_1 v u^2 - k_2 u^3 - k_3 u + k_4 w \tag{6.8}$$

where $k_1, k_2$, $k_3$ and $k_4$ are reaction rate constants and $v$ and $w$ are (fixed) concentrations of species $A$ and $B$.

In the problem studied by Fisher and by Kolmogorov with coworkers, the non-
populated state with $u = 0$ was absolutely unstable: even a vanishingly small initial
infection was enough to trigger the transition. In contrast to this, in chemical reactions
and combustion processes the function $f(u)$ often has the characteristic shape shown
in Fig. 6.2. Thus, the medium has two uniform stable states $u_1$ and $u_3$ and an unstable
state $u_2$.

Traveling fronts in bistable media represent waves of transitions between two
linearly stable states, such as $u_1$ and $u_3$ in Fig. 6.2. These fronts propagate at a
constant velocity and preserve their profile. However, their velocity is not given by
a simple Eq. (6.1) and the direction of propagation can become reversed if the form
of the function $f(u)$ is changed.

The direction of propagation is determined by the sign of the integral

$$Q = \int_{u_1}^{u_3} f(u)du \tag{6.9}$$

that corresponds to the difference of the shaded areas above and below the horizontal
axis in Fig. 6.2. If $Q > 0$, the front represents a wave of transition from the state

$u_1$ into the state $u_3$; the propagation direction is reversed if $Q < 0$. At $Q = 0$, a standing front is observed. Near the reversal point, the propagation velocity can be expressed as $V = aQ$ where $a$ is some coefficient.

The condition $Q = 0$ is similar to the Maxwell condition for the coexistence of two phases in equilibrium physical systems. Indeed, the states $u_1$ and $u_3$ can be viewed as corresponding to two different phases, so that the front represents an interface between them. When $Q = 0$, the two phases coexist. When $Q > 0$, the interface moves into the region occupied by the phase $u_1$ so that it becomes replaced by the phase $u_3$. Hence, the analogy with the crystallization waves, proposed by Ostwald, becomes plausible. It should be however stressed that this analogy is only formal. While thermodynamic properties determine the equilibrium phase states of the matter, such as vapour or liquid, steady states of a chemical bistable system are of a purely kinetic origin and they correspond to non-equilibrium conditions.

To initiate a spreading front in a bistable medium, initial perturbations should be sufficiently strong. Similar to equilibrium first-order phase transitions, a critical nucleus exists. Only if the size of a perturbation exceeds that of the critical nucleus, it would grow and the transition to a different state would take place. Smaller perturbations shrink and disappear after a while.

In combustion, traveling waves represent propagating fire (flame) fronts. Before the front, the combustion is absent and the medium is at the ambient temperature. After its passage, the fire sets on and the temperature is high. To initiate the fire, a sufficiently large region—a "critical nucleus"—must be initially heated. Under unfavorable conditions, i.e. when $Q < 0$, the front propagation velocity is negative: the fire retreats and quenching takes place.

The motivation for the original 1906 study by Luther [1] came, as already noted, from observations of electrical pulse propagation along the nerves. Apart from the similarities between traveling electric pulses and propagating chemical fronts, there is however also an important difference, and Luther was aware of it. After a front passage, the chemical medium undergoes a transition into a new steady state and remains to stay in this state. In contrast, the initial properties of a nerve are recovered after some time and a new pulse can propagate again along it. Thus, the nerve is *excitable*—many pulses can be excited one after another and may then propagate.

Already in the beginning of the 20th century, it was also known that cardiac tissue is excitable and electric waves can propagate through it. About once per second, such a wave is initiated in a certain region of the heart, it spreads concentrically and induces contractions of the cardiac muscle, producing the heart beat. However, under pathological conditions, the normal beat can be replaced by more complex propagation regimes.

The American mathematician Norbert Wiener, the founder of cybernetics, wanted to explain how complex wave patterns in the heart can arise. Together with his friend, a Mexican cardiologist Arthuro Rosenblueth, he published [8] in 1946 in *Archivos del Instituto de Cardiologia de Mexico* an article with the title "The mathematical formulation of the problem of conduction of impulses in a network of connected excitable elements, specifically in cardiac muscle". This publication has determined to a large extent the subsequent history of theoretical and experimental research on wave patterns in excitable media.

Wiener and Rosenblueth gave a clear description of excitable media:

> Conduction in nervous tissue resembles that in striated cardiac muscle. The laws which apply to the muscle fibers are also applicable to the nerve fibers. In both instances, the propagation is active, with energy supplied locally. In both cases, an impulse travels with a nearly uniform velocity. In both cases the excitation and transmission are all-or-none and do not allow for impulses of varying degrees of strength. In both cases activity is followed by an inexcitable period of definite duration, the absolutely refractory period; and this stage is followed in turn by a relatively refractory period, during which the tissues have subnormal excitability.

The aim of the authors was to explain, beyond normal heat beating, two other propagation regimes, found not only in the heart, but also in neural systems:

> If a turtle's ventricle is overdistended or stimulated with rapidly repeated electric shocks, a continuous cyclical wave appears, running along a definite closed path. In this condition, unlike physiological beating, the whole muscle is never at rest (in diastole). Similar activity has been observed in human auricles. The conditions which initiate this activity are not known. The phenomenon is called flutter.

> A type of nervous conduction quite like cardiac flutter is well known in the umbrella disc of coelenterates. Appropriate stimuli to the quasi-circular nerve net of a medusa or an anemone will start a circulating wave which may keep going around for many hours. Although neither the anatomical structure nor the physiological organization of this network is adequately known, the analogy to flutter in cardiac muscle is striking.

> The third type of cardiac contraction and conduction is known as fibrillation. Rapidly repeated stimulation of the mammalian ventricle will again set up circulating waves, as in flutter. These waves, however, do not seem to progress in an orderly fashion but rather to follow random paths. Several waves can progress simultaneously. The ventricle looks like a quivering mass of worms.

The flutter corresponds to a pathological condition of arrhythmia, whereas fibrillation, if not stopped, leads to instantaneous cardiac death. In the following chapters, we will show that similar wave propagation regimes are possible and have been observed in chemical excitable media as well.

At that time, the mechanisms underlying propagation of pulses in nerves and in the cardiac tissue were not known. A realistic model for electrical processes in the nerves was proposed (see below) only in 1952. The model of front propagation considered by Fisher and by Kolmogorov with coworkers in 1937 was formulated in terms of genetics and therefore its relevance for neural phenomena was not apparent (and it was not widely known).

Therefore, Wiener and Rosenblueth have built their model in the algorithmic (or automata) form. According to them, the system consists of locally connected excitable elements. Each of these elements can be found in the *rest*, *excitation* or *refractory* (i.e., recovery) states. The arrival of an external activation at an element in the rest state triggers its transition into the state of excitation. The element stays a short time $\tau_e$ in the excited state (in the original model, this duration was assumed to be vanishingly short) and then goes into the refractory state of a fixed duration $\tau_r$. While in this state, the element does not respond to external activation. Only when the refractory period has expired and the element has returned to the original rest state, it can be again externally activated starting a new excitation cycle.

An element that is in the excited state activates all neighbouring elements that are currently in the state of rest. Thus, in a linear chain of elements, an excitation pulse will propagate. At its front, the elements are excited, whereas they are in the refractory state in its back. The pulse propagates at a constant velocity $V$. The width of the traveling pulse is $V(\tau_e + \tau_r)$.

If the elements are very small, their locally connected network can be approximately treated as a continuous excitable medium with three discrete local states. Excitation waves can propagate through it. These waves may have complex shapes, but the propagation law is simple: Each small segment of the wave moves in its normal direction at the constant velocity $V$, independent of other elements (and thus this is similar to the situation in geometrical optics where light propagates at a constant velocity along the rays). To complete the definition, the law of motion of excitation waves at the boundaries of the medium had to be specified. It was assumed that a wave cannot separate from the boundary and it propagates orthogonally along it.

It follows from the model that, if two excitation pulses collide, they annihilate. Indeed, the excited elements in the collision zone are then surrounded by the elements in the refractory state. Therefore, they cannot activate other elements and the excitation dies out. Consequently, when traveling excitation waves collide, they annihilate.

A wealth of wave propagation phenomena in excitable media can be already described by the Wiener-Rosenblueth model. Suppose, for example, that an element in the medium is repeatedly externally excited at regular time intervals. Then it will give rise to concentric excitation waves that spread out away of it. Thus, this element will act as a *pacemaker*. In the heart, a group of the cells near the sinus node are in the oscillatory regime. This cell group acts as an intrinsic pacemaker and generates the physiological heat beat.

Moreover, interactions between the pacemakers can also be analyzed (but this was not done by Wiener and Rosenblueth). Let us consider two pacemakers with different frequencies $\omega_1$ and $\omega_2$ (with $\omega_1 > \omega_2$) whose centers are separated by distance $L$. Because the waves emanated from the pacemakers annihilate when they collide and the wave sequence sent by the first pacemaker is more rapid, the collision zone would gradually shift in the course of time, as illustrated in Fig. 6.3. Thus, after some time the activity of the slow pacemaker will be suppressed.

This result can be generalized to a system including a set of pacemakers with various frequencies. After sufficiently long time, only the most rapid pacemaker will survive and entrain the entire medium.

The effect is naturally employed in the heart to ensure high reliability of its functioning. The sinus node actually contains several groups of oscillatory cells with different periods. Under normal conditions, the most rapid of them dominates and sets the heart rhythm. If, however, the heart tissue is locally damaged and the most rapid group ceases to operate, the rhythm generation is automatically taken over by the next group which has however a lower frequency and this can be repeated several times.

**Fig. 6.3** Competition
between pacemakers. The
slower pacemaker with
period $T = 5$ becomes
suppressed by the rapid
pacemaker with period
$T = 1$

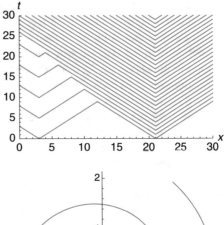

**Fig. 6.4** A spiral wave
rotating around an obstacle.
The pitch of the spiral is
equal to the perimeter of the
obstacle

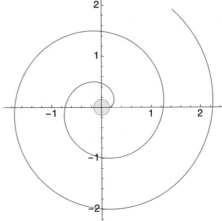

Another kind of phenomena in excitable media involves circulation of waves. In a narrow ring, an excitation wave can indefinitely circulate and the period is $T = 2\pi R/V$ where $R$ is the radius of the ring. Suppose now that we start to expand the ring and transform it into a wide annulus by keeping constant the radius of the inner hole. Obviously, the wave will continue to circulate, but its front should become curved. Indeed, if the wave were rotating as a straight ray, its outer parts had to move faster than the inner parts. But this is not possible: all segments of the wave propagate with the same velocity $V$ in the direction normal to them. Thus, the outer segments will retard and the wave will approach a spiral shape.

By continuing expansion of the ring, we come to the problem of wave circulation around a hole (or an obstacle) in an infinite medium that has been analyzed by Wiener and Rosenblueth. Their result was that the wave would form a rigidly rotating spiral representing the *involute* of the obstacle (Fig. 6.4) and having the rotation period $T = 2\pi R/V$.

An involute of a circle is a curve that was first considered by Christiaan Huygens. It can be rationalized in a simple way. Suppose that a wire is wound around a cylinder and a pencil it attached to its open end. If we start to unwind the wire while always keeping it straight, the trace left by the pencil on the paper will be the involute of

**Fig. 6.5** Spiral waves
around two obstacles.
Reproduced from [8]

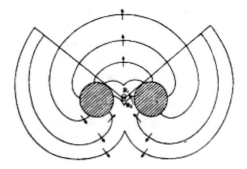

**Fig. 6.5** Spiral waves around two obstacles. Reproduced from [8]

the cylinder. Its property is that its pitch, i.e. the distance between two subsequent turns, is equal to the perimeter of the cycle. For example, the radius of the circle in Fig. 6.4 is $R = 1/2\pi$ so that its perimeter is of length one. Therefore, the separation between the turns is also equal to one.

If the medium contains several obstacles, waves can circulate around all of them. In Fig. 6.5 that we reproduce from the original article by Wiener and Rosenblueth, two spiral waves rotating in different directions around two obstacles are shown. Note that, as always, the waves annihilate when they collide.

Thus, any obstacle that pins a circulating excitation wave will, at a large separation from it, look like a source that sends the waves with the wavelength $\lambda = L$ and period $T = \lambda/V = L/T$, where $L$ is the perimeter of the obstacle. Note that this result holds even if an obstacle is not circular, but has a more complex shape.

Circulation of waves around obstacles, that can represent blood vessels or small regions of cardiac tissue damaged after an infarction, allows to explain the phenomenon of flutter. As for fibrillation, the understanding of this phenomenon was reached only much later (and we shall briefly discuss it in this chapter too).

Not every obstacle can maintain a rotating spiral wave. Indeed, if the perimeter $L$ of an obstacle is shorter than the pulse width $V(\tau_e + \tau_r)$, the wave cannot propagate around it because, when the excitation front returns after a full rotation, it finds the medium still in the refractory state. This implies that the radius of the smallest hole still supporting a spiral wave is $R_c = V(\tau_e + \tau_r)/2\pi$.

Wiener and Rosenblueth believed that excitation waves could only circulate around the obstacles. However, just two years later, O. Selfridge showed [9] that freely rotating spiral waves are also possible. His arguments were simple: Let us take a minimal obstacle of radius $R_c$ and start to deform it, while keeping the excitation wave rotating around it. The deformation consists in stretching the obstacle in one direction and compressing it in the other direction, so that the perimeter of the obstacle is not changed. If we continue such a deformation, we would eventually arrive at a situation where the obstacle is just a line of length $\pi R_c$, or half of the pulse width. Obviously, the wave can still circulate around this line obstacle, going up along its left side and down along the right side. But is the obstacle then needed at all? When the excitation wave moves up along the left side, it cannot anyway spread to the right because there are no elements there that are in the state of rest. Thus, we can remove the obstacle and obtain a freely rotating spiral wave.

In 1952, British scientists Alan Hodgkin and Andrew Huxley have discovered [10] the mechanism of pulse propagation along the nerve and constructed a mathematical model for such mechanism (they have received in 1963 a Nobel prize for their work). The mechanism did not involve chemical reactions or diffusion phenomena. Instead, the pulse propagation was of purely electric nature and it was controlled by opening and closing ion channels in the biological membrane of the nerve. The energy was continuously supplied due to a difference of ion concentrations inside and outside of the nerve. The energy supply allowed persistent propagation of the pulse.

The Hodgkin-Huxley equations are complicated and several kinds of ion channels are taken into account in them. Trying to simulate these equations on an analog computer, Richard FitzHugh has constructed [11] in 1961 a simplified version of them. One year later, the same simplified equations were proposed by J. Nagumo et al. [12]. The FitzHugh-Nagumo model is

$$\frac{\partial u}{\partial t} = u - \frac{1}{3}u^3 - v + \frac{\partial^2 u}{\partial x^2} \tag{6.10}$$

$$\tau \frac{\partial v}{\partial t} = u + a - bv \tag{6.11}$$

Here, certain time and coordinate units are used and the variables $u$ and $v$ are rescaled, so that only four dimensionless parameters $a$, $b$ and $\tau$ remain. The parameter $\tau$ defines the time scale for the variable $v$ and typically $\tau \gg 1$. The characteristic time for variable $u$ is of order unity.

In the original neurophysiological formulation, the variable $u$ in the FitzHugh-Nagumo model corresponded to the local potential along the nerve and the last term in Eq. (6.10) took into account electrical coupling between the elements in the overdamped regime. The second variable $v$ was used to take into account the combined effects of ion channels. However, these equations can also be interpreted in other terms.

The variable $u$ may also correspond to the chemical concentration of a species that can reproduce and diffuse. Indeed, the Eq. (6.10) for $u$ coincides with with the equation (6.4) where the function $f(u, v) = u - u^3/3 - v$ has the characteristic form shown in Fig. 6.2. When $v$ is increased, the growth of $u$ is suppressed and therefore the variable $v$ corresponds to an inhibitor species. Looking at Eq. (6.11), we can furthermore see then that the inhibitor kinetics is determined by the function $g(u, v) = (u + a - bv)/\tau$. The inhibitor is produced by the species $u$ and also undergoes decay; it does not diffuse. Note that variables $u$ and $v$ in these equations may take negative values and therefore they should rather represent deviations of the concentrations from some reference levels.

Figure 6.6 shows a traveling excitation pulse in the FitzHugh-Nagumo model. Before the arrival of the pulse, the system is in the steady state $(u_1, v_1)$ that is stable

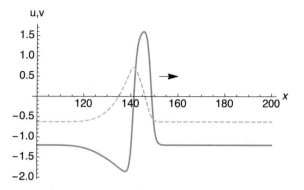

**Fig. 6.6** A traveling pulse in the FitzHugh-Nagumo model. Profiles of the activator (*solid*) and inhibitor (*dashed*) distributions are shown. The parameters are $a = 0.7, b = 0.8, \tau = 13$

with respect to sufficiently small perturbations.The front is similar to the transition wave already considered by Luther. At the front, the concentration of the reproducing species rapidly increases and saturates, whereas the inhibitor concentration remains approximately constant because the characteristic time scale for the inhibitor is long. Later on, however, slow growth of the inhibitor concentration begins. Eventually, the inhibitor concentration becomes so large that the state with high reproduction of the autocatalytic species cannot be maintained and the system returns to its original state.

In Chaps. 7 and 8, examples of actual chemical reactions with excitable kinetics will be given. Essentially, all of them belong to the class of the activator-inhibitor systems introduced by Turing and described by Eqs. (4.1) and (4.2). However, excitability is found at the conditions that are different from those assumed by Turing. For Turing patterns, diffusion of the inhibitor species should be fast. In contrast to this, the inhibitor should diffuse only slowly or even be immobile for excitation waves. Moreover, its kinetics should be slow as compared to that of the activator. But the most important difference is that, in the excitable case, the uniform state is stable with respect to small perturbations whereas it gets unstable after the Turing bifurcation.

The activator-inhibitor systems can also be bistable, so that two different uniform stationary states $(u_1, v_1)$ and $(u_2, v_2)$ coexist. The wave propagation phenomena in such two-component systems are however more rich than in the one-component bistable systems considered above. Two kinds of fronts, that correspond to a transition from $(u_1, v_1)$ to $(u_2, v_2)$ or from $(u_2, v_2)$ to $(u_1, v_1)$, can propagate through such a medium for the same parameter values. This means that *both* stationary states are metastable and, by applying an appropriate perturbation, a transition to another state can be initiated, in contrast to the former one-component case [13].

The two fronts have different propagation velocities. Suppose that the velocity $V_1$ of the front 1 that transfers the state $(u_1, v_1)$ into the state $(u_2, v_2)$ is smaller than the velocity $V_2$ of the front 2 that transfers the state $(u_2, v_2)$ back to $(u_1, v_1)$. What will happen if the two fronts run after another, with the front 2 after the front 1? Obviously, the system will find itself in the same state $(u_1, v_1)$ after the propagation of the two of them. Because the second front, implementing the reverse transition, moves at a higher velocity than the first front, it will eventually run into it. As a result of the collision, the two fronts can annihilate.

It is however also possible that a structure representing the *bound state* of two fronts and traveling at a constant velocity (close to $V_1$) will be formed. The states of the medium before and after it will be the same and therefore this structure will effectively represent an excitation wave, similar to the traveling pulse shown in Fig. 6.6. Thus, excitation waves can also be observed in bistable activator-inhibitor systems [13].

Numerous simulations and analytical studies for various models of excitable media are available, but, in this book, we cannot provide a detailed review of them. Essentially, their results agree with those of the axiomatic Wiener-Rosenblueth model. However, the propagation velocity of excitation waves is not constant. If the next pulse propagates through a medium where, shortly before, another pulse has propagated, the velocity is decreased because the inhibitor concentration is still enhanced. Moreover, the propagation velocity depends approximately as $V = V_0 - DK$ on the geometric local curvature $K$ of the wave (here $V_0$ is the velocity of the flat front and the coefficient $D$ is close to the diffusion constant of the activator species).

There was nonetheless one important disagreement between predictions of the theory based on the Wiener-Rosenblueth model and numerical simulations of reaction-diffusion systems. The simulations showed that spirals could have very large cores around which the excitation wave circulates. The rotation period of such spirals was much larger than the refractory time. The core region was never visited by the waves and remained in the state of rest.

Figure 6.7 shows an example of the simulations where such spiral waves were formed. As the initial condition, a broken excitation wave was used. It was obtained by erasing the upper part of the wave, i.e. by resetting the elements to the rest state. Thus, an excitation wave with a free tip was prepared. The subsequent evolution of this broken wave was followed in the simulations. When excitability was relatively low, the wave started to contract (Fig. 6.7a). At some critical excitability, a broken wave that propagated neither contracting, nor growing was seen (Fig. 6.7b). At a higher excitability, the wave started to grow and to curl (Fig. 6.7c). When the excitability was further increased, a freely rotating spiral wave with a large core was formed (Fig. 6.7d). Note that the central core region never became excited and the medium there was in the state of rest. When simulations with a medium of a larger size were performed, a spiral wave with an even larger core could be seen for the parameters corresponding to Fig. 6.7c.

To reproduce such behavior, an extension of the Wiener-Rosenblueth model was proposed [14, 15]. In this extension, excitation waves were modeled by curves that could have open ends. Each element of the curve propagated in its normal direction with the velocity $V = V_0 - DK$ that depended on its curvature. Additionally, the curve could grow or contract in the tangential direction at its open end. The growth velocity $G$ depended on the curvature $K_0$ at the open end as $G = G_0 - \gamma K_0$ where $G_0$ is the growth velocity for a broken flat excitation wave.

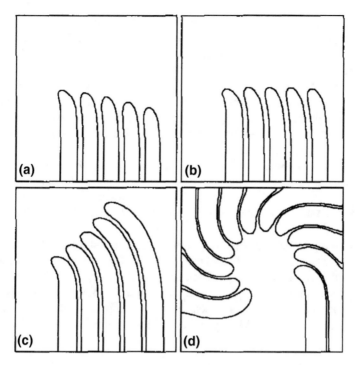

**Fig. 6.7** Evolution of a broken excitation wave in a reaction-diffusion model of an excitable medium. Contours of constant activator concentration are shown at subsequent time moments for four different **a–d** excitabilities of the medium. Reproduced from [14]

**Fig. 6.8** Formation of a freely rotating spiral wave in the extended Wiener-Rosenblueth model. Subsequent positions of the curve representing the excitation wave are given. The *dashed line* shows the trajectory of the wave tip. Reproduced from [15]

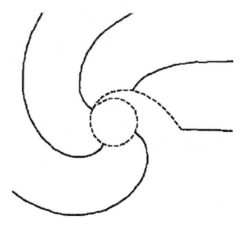

By running numerical simulations with this model, the behavior shown in Fig. 6.7 could be reproduced. The simulation in Fig. 6.8 was started with a broken flat wave and a freely rotating spiral with the core not visited by the waves was formed. If the parameter $G_0$ was chosen negative, contracting broken flat waves, like in Fig. 6.7a, could be observed. The curvature of the final spiral wave at its tip was equal to $K_0 = G_0/\gamma$ so that $G = 0$. Thus, this curved wave neither grew nor contracted in its tangential direction at its tip. It rigidly rotated around a circular core and moved orthogonally to it. This condition allowed to determine the rotation frequency of the spiral wave and the radius of its core [15].

Another class of systems where traveling waves can be observed are oscillatory media. In Chap. 5, we have already considered chemical oscillations, but assumed that the reactors were well stirred, so that spatial inhomogeneities and diffusion effects do not take place. Below the diffusion effects in chemical oscillatory media will be included.

As an example, systems near a Hopf bifurcation where the oscillations are almost harmonic and have small amplitudes can be chosen. For such systems, details of the kinetic models do not play a significant role and the analysis can be performed in a unified way. Similar to the previous description in Chap. 5, a complex oscillation amplitude $\eta$ can be introduced which depends now both on time and on spatial coordinates, $\eta = \eta(x, t)$. The dynamics of this amplitude is determined by the equation

$$\frac{\partial \eta}{\partial t} = \gamma \eta + i \omega \eta - (\alpha + i \beta) |\eta|^2 \eta + (\mu + i \nu) \nabla^2 \eta \qquad (6.12)$$

where the last term is due to the diffusion effects. The local phase $\phi(x, t)$ and magnitude $\rho(x, t)$ of oscillations can be introduced as $\eta(x, t) = \rho(x, t) \exp(i \phi(x, t))$.

The coefficient $\mu$ in Eq. (6.12) is always positive, whereas the coefficient $\nu$ can have any sign. If diffusion constants of all chemical species are equal, $\nu = 0$. Therefore, this coefficient is determined by *differences* in the diffusion constants. For an activator-inhibitor reaction-diffusion system (4.1)–(4.2), we have $\nu \propto (D_u - D_v)$.

The model (6.12) is known as the *complex Ginzburg–Landau equation*. Concerning its name, some comments have to be made. In 1950, Russian physicists Vitaly Ginzburg and Lev Landau constructed [16] a similar equation for the complex order parameter in quantum superconducting systems near an equilibrium phase transition. In this Ginzburg–Landau equation, there were however no imaginary terms, i.e. $\omega = \beta = \nu = 0$. The Eq. (6.12) was proposed in 1975 by Yoshiki Kuramoto and Toshio Tsuzuki [17] for classical chemical systems, but traditionally it is nonetheless viewed as an extension of the original theory by Ginzburg and Landau. In the context of fluid convection problems, Eq. (6.12) was derived in 1969 by Alan Newell and John Whitehead [18].

The complex Ginzburg–Landau equation accounts for traveling plane waves, $\eta(x, t) = \rho_k \exp[i(kx - \omega_k t)]$. The magnitude $\rho_k$ of a wave with the wavenumber $k$ and spatial period $\lambda = 2\pi/k$ is $\rho_k = (\gamma - \mu k^2)/\alpha$ and its frequency $\omega_k$ is given by $\omega_k = \omega - \beta \rho_k - \nu k^2$. The frequency can increase or decrease with the

 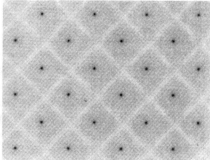

**Fig. 6.9** A lattice formed by rotating spiral waves. Numerical simulation for the complex Ginzburg–Landau equation with the parameters $\gamma = \alpha = \mu = 1$, $\beta = 0.7$, and $\nu = 0$. Snapshots of the real part (*left*) and the modulus (*right*) of the complex oscillation amplitude are shown. Reproduced with permission from [20]

wavenumber, so that the waves have positive or negative dispersion, depending on the sign of the coefficient $\nu$. Note that the wave amplitude $\rho_k$ is not arbitrary, but uniquely determined by the wavenumber $k$. Only the waves with sufficiently small wavenumbers $k < \sqrt{\gamma/\alpha}$ can exist.

Thus, in the case of the Hopf bifurcation, first the stationary state becomes unstable with respect to uniform oscillations and only after that, above the bifurcation point, traveling waves can propagate. There is however also a different kind of instability, known as the *wave bifurcation*. Then, a uniform stationary state becomes first unstable with respect to spontaneous emergence of the waves with a definite wavenumber $k_c$ and frequency $\omega_c$. The effect is analogous to the Turing bifurcation, with the only difference that the critical mode is a traveling, not a stationary, wave. In his 1952 article, Turing described this other bifurcation too. He has shown [19] that, in contrast to the bifircation leading to stationary periodic patterns that exists already for a system with two reacting species, at least three reacting species are needed for the wave bifurcation to occur.

Spiral waves are also possible in complex Ginzburg-Landau model. Figure 6.9 shows, as an example, a stable lattice formed by spiral waves with opposite rotation directions. The real part of the complex oscillation amplitude Re $\eta$ and the magnitude $\rho$ are displayed.

If we choose a circuit surrounding the center of a spiral wave and go around this circuit, the oscillation phase will increase or decrease by $2\pi$ after one turn, depending on whether the wave is rotating clock- or anticlockwise. If we shrink this circuit, the phase should always change by $2\pi$ after a turn even if the circuit becomes vanishingly small. But what is then the value of the phase exactly in the center? It cannot be defined.

The solution to this apparent paradox is that the oscillation amplitude $\rho$ vanishes in the spiral center. Hence, there are no oscillations in this point and the phase cannot be assigned. Indeed, the modulus $\rho$ of the complex oscillation amplitude decreases as the center of a spiral wave is approached and becomes zero exactly at this point. Hence, any spiral corresponds to an *amplitude defect*. Such amplitude defects located in the spiral wave centers are clearly seen in the right part of Fig. 6.9.

The spiral waves represent autonomous wave sources. All single spirals have the same frequency and shape, determined solely by the parameters of an oscillatory medium. Only the location of a spiral wave and its rotation direction are determined by initial conditions and can therefore change.

The uniform oscillations in an oscillatory medium can be unstable with respect to spatial modulation and generation of amplitude defects. For the complex Ginzburg–Landau equation (6.12), such *Benjamin-Fair instability* takes place if the condition

$$\mu + \frac{\beta}{\alpha}\nu < 0 \tag{6.13}$$

is satisfied. Because $\mu > 0$ and the coefficient $\alpha$ is also positive in the considered case of a supercritical Hopf bifurcation, this condition implies that the coefficients $\beta$ and $\nu$ should have opposite signs.

Note that the instability condition (6.13) also implies that $|\nu| > \alpha\mu/|\beta|$. Since, for the two-component activator-inhibitor systems, we have $|\nu| \propto (D_u - D_v)$, such instability can only develop if the difference in the diffusion constants of the activator and inhibitor species is sufficiently large. In this sense, the instability is similar to the Turing bifurcation considered in Chap. 4.

In contrast to the Turing bifurcation that leads to the emergence of stationary self-organized patterns, the Benjamin-Fair instability of uniform oscillations results in the development of complex wave regimes, known as *chemical turbulence* and first considered by Kuramoto [21]. Pairs of spiral wave with opposite rotation directions and amplitude defects in their centers spontaneously emerge. Such spiral waves give rise to further spirals and thus a reproduction cascade of amplitude defects takes place. As a result, the medium becomes filled with multiple wave fragments that move in an irregular way (Fig. 6.10). This regime can be characterized as spatiotemporal chaos and represents an analog of chaotic oscillations described at the end of the previous chapter. It is also similar to hydrodynamic turbulence in stirred fluids.

Chemical turbulence is also possible in excitable media. Its examples will be given in the next two chapters where the Belousov–Zhabotinky reaction in aqueous solutions and systems with surface chemical reactions are discussed. It proceeds through the reproduction cascade of rotating spiral waves. In this regime, numerous excitation wave fragments, colliding and moving along random paths are observed. In the cardiac tissue it corresponds to the lethal condition of heart fibrillation known already by Wiener, who, however, gave no explanation to it. The medium looks then indeed like a "quivering mass of worms" [8].

So far, patterns of traveling waves in one- and two-dimensional media were considered in this chapter. However, three-dimensional oscillatory or excitable media also exist. Homogeneous chemical reactions in bulk reactors obviously provide one example. The cardiac muscle has also a certain thickness and therefore three-dimensional effects can be important for it.

**Fig. 6.10** Chemical
turbulence. Numerical
simulation of the complex
Ginzburg–Landau equation.
The real part of the complex
oscillation amplitude is
displayed. (M. Hildebrand
and A. S. Mikhailov,
unpublished)

**Fig. 6.11**   A scroll ring. The
*dashed curve* indicates the
filament

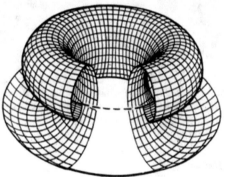

If we take a planar spiral wave and extend it in the vertical direction, a *scroll wave* is obtained. In its center, a thin cylindrical filament (corresponding to the core of a spiral wave) will exist. This pattern can be deformed in various ways, so that the filament becomes bent. Particularly, a scroll ring schematically shown in Fig. 6.11 can develop. Such wave patterns were first described [22] by Arthur Winfree in 1973.

Under certain conditions, the filament of a scroll wave can become unstable [23, 24] with respect to bending and stretching (formally, their tension becomes negative). As a result, a scroll ring is transformed into an irregular wave pattern (Fig. 6.12) and a different kind of chemical turbulence, characteristic only for three-dimensional excitable media, is obtained. Winfree argued [25] that fibrillation can sometimes set on without multiple wave breakups and the negative-tension 3D instability provides an alternative mechanism for this [26].

In the next two chapters, experimental observations of traveling waves in different chemical systems will be presented and discussed.

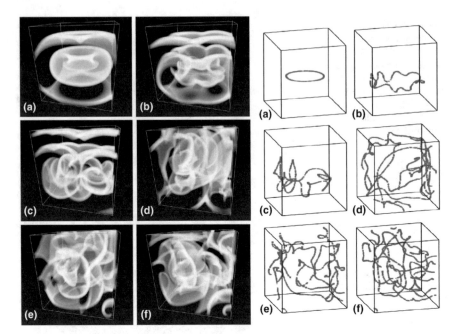

**Fig. 6.12** Development of chemical turbulence through the negative-tension instability of a scroll wave. Subsequent snapshots **a–f** from numerical simulations for a three-dimensional excitable medium. Distributions of the activator concentration (*left*) and the filaments of scroll waves (*right*) are shown. Reproduced from [27]

# References

1. R. Luther, Z. Elektrochemie 12, 596 (1906) [English translation: J. Chem. Education **64**, 740 (1987)]
2. K. Showalter, J. Tyson, J. Chem. Educ. **64**, 742 (1987)
3. A.J. Lotka, J. Phys. Chem. **14**, 271 (1910)
4. P. Verhulst, Correspondance Mathématique et Physique **10**, 113 (1838)
5. R.A. Fisher, Ann. Eugen. **7**, 355 (1937)
6. A. Kolmogorov, I. Petrovsky, N. Piskunov, Bull. Mosc. Univ. Ser. A Math. Mech. **1**, 1 (1937)
7. F. Schlögl, Z. Phys. **253**, 147 (1972)
8. A. Wiener, A. Rosenblueth, Arch. Inst. Cardiol. Mex. **16**, 205 (1946)
9. O. Selfridge, Arch. Inst. Cardiol. Mex. **18**, 177 (1948)
10. A.L. Hodgkin, A.F. Huxley, J. Physiol. **117**, 500 (1952)
11. R. FitzHugh, Biophys. J. **1**, 445 (1961)
12. J. Nagumo, S. Arimoto, S. Yoshizawa, Proc. IRE **50**, 2061 (1962)
13. L.S. Polak, A.S. Mikhailov *Samoorganizatiya v Neravnovesnykh Fiziko-Chimicheskikh Systemakh* (Self-organization in Nonequilibrium Physical-Chemical Systems) (Moscow, Nauka, 1983)
14. V.A. Davydov, A.S. Mikhailov, V.S. Zykov "Kinematical theory of autowave patterns in excitable media", in *Nonlinear Waves in Active Media*, ed. Yu. Engelbrecht (Springer, Berlin 1989), pp. 38–51
15. A.S. Mikhailov, V.A. Davydov, V.S. Zykov, Physica D **70**, 1 (1994)

16. V.L. Ginzburg, L.D. Landau, Zh. Eksp. Teor. Fiz. **20**, 1064 (1950)
17. Y. Kuramoto, T. Tsuzuki, Prog. Theor. Phys. **54**, 687 (1975)
18. A.C. Newell, J.A. Whitehead, J. Fluid Mech. **38**, 279 (1969)
19. A.M. Turing, Philos. Trans. R. Soc. Lond. B **237**, 37 (1952)
20. I.S. Aronson, L. Kramer, A. Weber, Phys. Rev. E **48**, R9 (1993)
21. Y. Kuramoto, *Chemical Oscillations, Waves and Turbulence* (Springer, Berlin, 1984)
22. A.T. Winfree, Science **181**, 937 (1973)
23. P.K. Brazhnik, V.A. Davydov, V.S. Zykov, A.S. Mikhailov, Sov. Phys. JETP **66**, 984 (1987)
24. V.N. Biktashev, A.V. Holden, H. Zhang, Philos. Trans. R. Soc. Lond. Ser. A **347**, 611 (1994)
25. A.T. Winfree, Science **266**, 1003 (1994)
26. S. Alonso, F. Sagues, A.S. Mikhailov, Science **299**, 1722 (2003)
27. S. Alonso, R. Kähler, A.S. Mikhailov, F. Sagues, Phys. Rev. E **70**, 056201 (2004)

# Chapter 7
# The Belousov–Zhabotinsky Reaction

The Belousov–Zhabotinsky (BZ) reaction is used extensively as a model chemical system in the studies of self-organization phenomena. Thus, it plays a role similar to that of *Drosophila*, or the fruit fly, in molecular genetics. It has not been the first oscillatory reaction discovered, and its full kinetic mechanism is quite complex and still not completely understood. Nonetheless, it can be easily prepared and run; it is also easy to control. The oscillation periods are in the range of tens of seconds and the wavelengths of spatial patterns are in the millimeter range, allowing for observations by the non-aided eye. While the reaction has no practical applications, there is extensive literature devoted to it.

Boris Belousov studied chemistry at the Swiss Federal Institute of Technology (ETH) in Zurich and returned to Moscow in 1916. He had mainly worked in the field of defence against chemical weapons and in radiation toxicology. He had also taught at the military Chemical Academy and retired in 1939 with the rank of a major-general. After that, he got a laboratory in the Biophysics Institute of the Ministry of Health in Moscow.

In 1950, Belousov was investigating the biochemical Krebs cycle that plays a central role in cell metabolism. He wanted to see whether a part of it could be catalyzed by free cerium ions instead of the protein-bound metal ions in the enzymes. With this aim, he has considered a chemical reaction of the cerium-catalyzed oxidation of citric acid by bromate in the sulphuric acid medium. To his surprise, the solution periodically changed its color from yellow to transparent and such oscillations could be seen for an hour or more. Intrigued, he had decided to explore this phenomenon in detail.

The yellow color of the solution was due to the cerous $Ce^{4+}$ ions, whereas the solution was transparent if the ceric $Ce^{3+}$ ions were prevailing in it. Hence, the state of the catalyst was apparently periodically changed. The typical oscillation period was about 80 s, but oscillations could be accelerated if concentrations of hydrogen or cerium ions were increased. By adding fresh citric acid or bromate, the oscillations could be prolonged. They became faster when temperature had been

© Springer International Publishing AG 2017
A.S. Mikhailov and G. Ertl, *Chemical Complexity*, The Frontiers Collection,
DOI 10.1007/978-3-319-57377-9_7

**Fig. 7.1** The first
mechanistic scheme of the
BZ reaction. Reproduced
from [6]

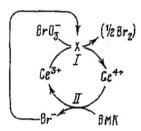

raised. Moreover, in a tube with the vertical concentration gradient, spatial patterns of stripes were observed. Belousov has taken photographs to document oscillations and the spatial patterns.

These results were described in the manuscript [1] that Belousov submitted in 1951 to the Soviet *Journal of General Chemistry*. His paper was rejected because, according to the referee, the results contradicted the Second Law of thermodynamics. In 1955, he submitted a revised version to another Soviet journal, *Kinetics and Catalysis*, and got a rejection again.

Meanwhile, the reaction became popular and the friends of Belousov convinced him to publish at least a brief account of it. Eventually, an abstract appeared [2] in the proceedings of the Biophysics Institute where Belousov worked. This short publication did not even include the reaction recipe. Only much later, when the reaction became broadly known, the original manuscript from the archive of Belousov was reproduced and its English translation appeared too [1]; the photographs had however been lost. Winfree has published in 1984 an article [3] about the discovery by Belousov, based on his manuscript and on the information he gathered from his colleagues.

In 1961, Anatol Zhabotinsky was at the beginning of his Ph.D course. Originally, he wanted to investigate glycolitic oscillations, but his supervisor S. Shnoll, who was familiar with Belousov, suggested that his reaction should be chosen instead. Zhabotinsky replaced the original citric acid with malonic acid to improve the optical contrast of color oscillations. He also analyzed, independently of the previous work by Belousov, the reaction mechanism. Only when his work was completed, he sent his first manuscript to Belousov and received in return the unpublished manuscript from him. Afterwards they were discussing the results per mail and on phone, but never met each other. After getting his Ph.D. degree, Zhabotinsky continued his investigations of the reaction and his major contribution was that he explored its rich pattern formation effects. In 1991, he moved to the Brandeis university in the US.

The BZ reaction combines processes of autocatalytic reproduction and product inhibition. Based on his own work [4, 5], Zhabotinsky gave [6] in 1991 the following description of its mechanism (see also Fig. 7.1):

> The reaction consists of two main parts: the autocatalytic oxidation of cerous ions by bromate and the reduction of ceric ions by malonic acid. Bromoderivatives of malonic acid are produced during the overall reaction. The ceric ion reduction is accompanied by the production of bromide ion from the bromoderivatives. Bromide ion is a strong inhibitor

of the autocatalytic oxidation because of its rapid reaction with the autocatalyst, which is presumably bromous acid or some oxibromine free radical.

An oscillatory cycle can be qualitatively described in the following way. Suppose that a sufficiently high ceric ion concentration is present in the system. Then, bromide ion will be produced rapidly and its concentration will also be high. As a result, autocatalytic oxidation is completely inhibited and the ceric ion concentration decreases due to its reduction by malonic acid. The bromide ion concentration decreases along with that of ceric ion. When $[Ce^{4+}]$ reaches its lower threshold, the bromide ion concentrations drops abruptly. The rapid autocatalytic oxidation starts and raises the ceric ion concentration. When this concentration reaches its upper threshold, $[Br^-]$ increases sharply, completely inhibiting the autocatalytic oxidation. The cycle then repeats.

The first paper on the BZ reaction outside of the Soviet Union was in 1967 by Hans Degn [7] who showed a possible important role of bromoderivatives of malonic acid, other than bromomalonic acid. In 1972, Richard Field, Endre Körös and Richard Noyes described [8] a detailed reaction mechanism.

The BZ reaction is complicated and its full description [9] includes 80 elementary steps whose rate constants are often not known. It is important, however, that oscillations are found inside a large region of reactant concentrations, extending over orders of magnitude along each axis [6]. Such robustness makes the reaction perfect for studies of oscillations and waves in chemical systems. When ferroin is used as the indicator, the role of cerium ions is taken by $Fe^{3+}$ and $Fe^{4+}$.

Because of its high complexity, no kinetic models of the BZ reaction that would have yielded quantitative agreement with the experiments are available. A broadly used mathematical description was proposed [10] by Field and Noyes at the University of Oregon. This *Oregonator model* is obtained by further simplification of the reaction scheme. It has the form

$$\frac{dX}{dt} = k_1 AY - k_2 XY + k_3 BX - 2k_4 X^2 \tag{7.1}$$

$$\frac{dY}{dt} = -k_1 AY - k_2 XY + fk_5 Z \tag{7.2}$$

$$\frac{dZ}{dt} = k_3 BX - k_5 Z \tag{7.3}$$

where $X = [HBrO_2]$, $Y = [Br^-]$, $Z = [Ce^{4+}]$, $A = [BrO_3^-]$ and $B = [CO_2(COOH)_2]$. The reactants A and B are normally present at high concentrations and they are assumed to be constant in the model; $f$ is a stochiometric factor.

The reaction scheme corresponding to Eqs. (7.1)–(7.3) is

$$A + Y \rightarrow X \tag{7.4}$$

$$X + Y \rightarrow P \tag{7.5}$$

$$B + X \rightarrow 2X + Z \tag{7.6}$$

$$2X \rightarrow Q \tag{7.7}$$

$$Z \rightarrow fY \tag{7.8}$$

All reactions are considered as irreversible in this scheme. The strongest simplification is in the last reaction step where a stochiometric factor $f$, that can take even fractional values, is phenomenologically introduced.

John Tyson and Paul Fife noticed [11] that typically the intermediate product $Y$ has fast kinetics and its concentration can be chosen as adjusting to instantaneous concentrations of $X$ and $Z$. Introducing dimensionless concentrations $u$ and $v$ of species $X$ and $Z$ and using certain time units, they arrived at a system with only two equations

$$\epsilon \frac{du}{dt} = u(1 - u) - \frac{bv(u - a)}{u + a} \tag{7.9}$$

$$\frac{dv}{dt} = u - v \tag{7.10}$$

where dimensionless parameters $\epsilon$, $a$ and $b$ represent combinations of reaction rate constants and conditions $\epsilon \ll 1$, $a \ll 1$ and $b \cong 1$ are satisfied.

Equations (7.9) and (7.10) are similar to the FitzHugh–Nagumo model (6.10)–(6.11). Moreover, they belong to the general class of activator-inhibitor systems that we have considered in Chaps. 4 and 5. The species $HBrO_2$ with the rescaled concentration $u$ is effectively autocatalytic and plays the role of an activator. Its reproduction is inhibited by the second species, i. e. by the cerous ions $Ce^{4+}$. The inhibitor is a product of the activator and, additionally, it is subject to decay. Because $\epsilon \ll 1$, the inhibitor concentration can change only slowly as compared to that of the activator.

Note that, similar to the FitzHugh–Nagumo model, this system can be oscillatory or excitable depending on the parameters $a$ and $b$. If reaction is not performed in a stirred reactor, diffusion terms should be included into Eqs. (7.9) and (7.10). Because molecular weights of the activator and inhibitor species in the BZ reaction do not vary greatly, their diffusion contants are of the same order of magnitude. In contrast to this, if the the FitzHugh–Nagumo Eqs. (6.10)–(6.11) are chosen to describe nerve propagation, there is no diffusion-like term in the second of them (because variable $v$ describes then a local property, i.e. the ion conductivity of the biomembrane). Despite this difference, Eqs. (7.9) and (7.10) support propagation of pulses in the excitable regime.

Already Ostwald and Luther tried to find physical or chemical processes that would yield persistent propagation of waves, similar to traveling electric pulses in the nerves. However, they could only show that traveling fronts exist. The discovery by Belousov allowed to identify a chemical system that has the same mathematical wave propagation properties as the nerves, although based on a completely different physical mechanism.

Along with the mechanistic studies, Zhabotinsky performed detailed investigations of the oscillations. With this aim, he built a flow reactor (CSTR) with constant infusion of bromide ions. He also used ultraviolet (UV) radiation to control the reaction course. Thus, oscillations with a more complex shape of bursts could be obtained [12] and entrainment of oscillations by periodic external forcing could be achieved [13].

Another direction pursued was to study spatial oscillating patterns and traveling waves. First, inhomogeneous oscillations in the tubes were considered, but they were apparently strongly influenced by convection effects. Therefore, experiments were then performed in thin solution layers in a Petri dish. Moreover, the ferroin oscillator was used, so that the solution color was changing from red to blue.

At the same time, Israel Gelfand in the Mathematics Department of the Moscow State University became interested in the problems of wave propagation in the brain and in the cardiac tissue. In the seminars of his group, the work [14] by Wiener and Rosenblueth was being actively discussed (the Russian translation of their article has been published in 1961 [15]). Zhabotinsky was in contact with some participants of these seminars and thus familiar with the concept of excitable media.

In 1970, first observations of pacemakers in the BZ reaction were reported by Zaikin and Zhabotinsky [16]. Initially, the oscillatory solution in the Petri dish was uniform, but after a while some spots developed inside it and these spots started to send concentric waves (Fig. 7.2). When such waves collided, annihilation was observed. In the course of time, slower pacemakers became eliminated and the fastest ones survived. This was explained by the arguments that we have already presented earlier in the previous chapter (see Fig. 6.3). The paper ended with a remark that "similar models have been applied to the explanation of impulse propagation in cardiac muscle" with the reference to the article by Wiener and Rosenblueth [14].

Later, Tyson and Fife have suggested [11] to describe the patterns of concentric waves as "target patterns" and this term became common afterwards. An example of a developing target pattern is shown in Fig. 7.3. As Zhabotinsky and Zaikin reported [17] in 1973, oscillation periods of different pacemakers varied more than twofold in the same experiment, they were however always shorter than the period of uniform oscillations. Target patterns were also observed in excitable media where the uniform stationary state is stable.

Most of the target patterns observed in the BZ reaction are caused by heterogeneities located in their centers. Such heterogeneities can increase the local oscillation frequency or make the medium locally oscillatory in the excitable case. There are however also indications that some of such patterns may be self-organized. It was

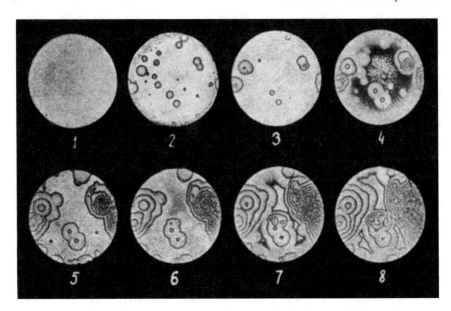

**Fig. 7.2** Multiple pacemakers in the BZ reaction. The snapshots are taken with the interval of 1 min. The diameter of the Petri dish is 100 mm. Reproduced with permission from [16]

shown [17] that the mean number of the pacemakers per unit area and the mean period of their oscillations depended on the chemical composition of the system, but were practically independent of incidental contaminations or the material from which the reactor was made. Such mean properties also remained similar after repeated filtration or when aluminium oxide powder was added. Moreover, a local inhomogeneity in the initial conditions, provided that it was strong enough, could create a pacemaker that persists afterwards. Later, C. Vidal and A. Pagola examined central regions of some target patterns by optical microscopy and they often could not see the presence of any contamination down to the micrometer scale [18].

Autonomous self-organized pacemakers were repeatedly reported in theoretical studies of three-component reaction diffusion models. However, this has never been done for realistic models of the BZ reaction. Note that, if pacemakers are self-organized, their frequencies should be uniquely determined by the properties of the active medium. Therefore, all such pacemakers should have the same frequency and thus coexist. This behaviour has not however been seen in the experiments with the BZ reaction.

Another important class of wave patterns in the BZ reaction are spiral waves. Such patterns were mentioned [19] by Zhabotinsky in 1970 and a photo showing spiral waves has been published [20] by him in 1971. Both these publications had however appeared in Russian and were not easily available. Outside of the USSR, these results had first appeared in press in 1973 [17].

In 1972, A. Winfree reported [21] his own observations of spiral waves, citing the previous work [19, 20]. His experiments were performed under excitable conditions

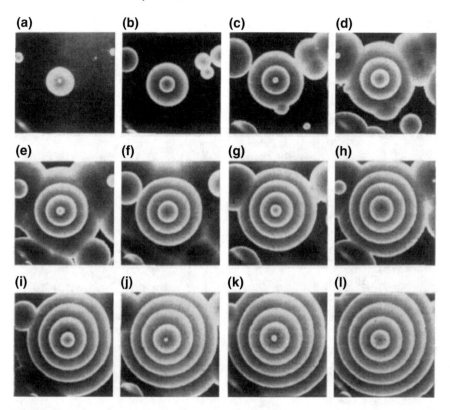

**Fig. 7.3** A target pattern in the BZ reaction. Subsequent photos taken at 30 s intervals. Reproduced with permission from [17]

and the medium could remain in the uniform state for a long time, although a few pacemakers could also emerge. In contrast to pacemakers, spiral waves could not spontaneously develop. To induce such patterns, Winfree had to locally erase the waves (propagating from a pacemaker), so that the waves with free ends had been obtained. Subsequently, propagating half-waves were winding around their free ends and spiral waves were thus obtained. The wavelength of the spiral wave was about 1 mm and the free end seemed to circulate around a core with the circumference of $1 \pm 0.5$ mm. Winfree had shown that spiral waves should have the form of an involute, repeating the results by Wiener and Rosenblueth.

The patterns of developing spiral waves in Fig. 7.4 are reproduced from the 1973 publication [17] by Zhabotinsky and Zaikin (no images of observed spiral waves were given by Winfree in his original 1972 report [21]). A circular wave is broken and, at each open end, the waves start to curle so that four spirals emerge. Note that that these rotating spiral waves coexist and thus their rotation frequencies should be the same.

Continuing his investigations of the BZ reaction in a Petri dish, Winfree has encountered [22] wave structures shown in the left panel of Fig. 7.5. These patterns

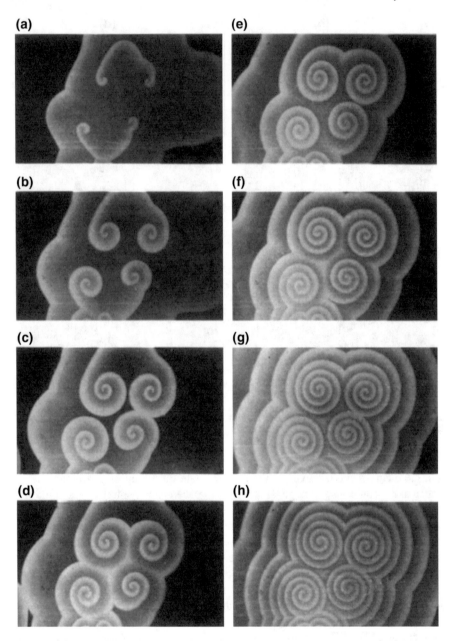

**Fig. 7.4** Spiral waves in the BZ reaction. The pattern develops after a break of a circular wave. The same intervals between snapshots (**a**–**h**) and the same initial concentrations of reactants as in Fig. 6.6. Reproduced with permission from [17]

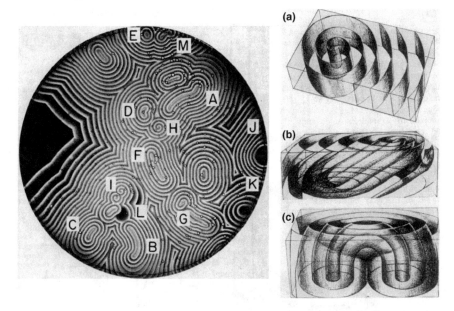

**Fig. 7.5** Scroll waves in the BZ reaction. *Left* Various visible wave structures in a thin layer (1.5 mm) of the reagent. *Right* Graphic interpretation of such apparent structures as different vertical projections of scroll waves. Reproduced with permission from [22]

were obtained starting with pacemakers and briefly shearing the solution to create crossed concentration gradients. Winfree has conjectured that such structures, seen on the surface of the dish, represent vertical projections (right panel in Fig. 7.5) of differently oriented scroll waves or rings (discussed by us in Chap. 5). To prove this, Winfree had designed an ingineous scheme:

> Waves propagate equally well in the homogeneous, porous, and relatively inert matrix provided by a Millipore filter. [...] Thus, a further test is to try three-dimensional reconstruction of scroll waves from serial section: waves propagating through stacked Millipores can be examined in cross-section (as though microtomed) by inducing waves from suitable initial conditions, letting them develop for several minutes (at least ten scroll rotations), then plunging and dispersing the stack into cold 3 percent perchloric acid. All wave patterns are fixed within about 1 second by the cold, the dilution of all inorganic ions, and formation of the insoluble ferrous phenanthroline perchlorite complex. Restacked in their original alignments, these filters reveal a diversity of wave patterns. Most of them can be described as bits and pieces of scroll waves such as Fig. 7.5, right, a–c. Occasionally, a complete scroll ring is found, such as Fig. 7.5, right, c.

Thus, first experimental observations of three-dimensional scroll wave patterns in the BZ reaction were made [22]. Later on, methods of optical tomography have been developed to achieve non-invasive observations of wave patterns in the 3D reagent [23].

Chaotic (irregular) oscillations for the BZ reaction were first observed in 1977 by K. Schmitz, K. Graziani and J. Hudson [24]. The experiment was performed using

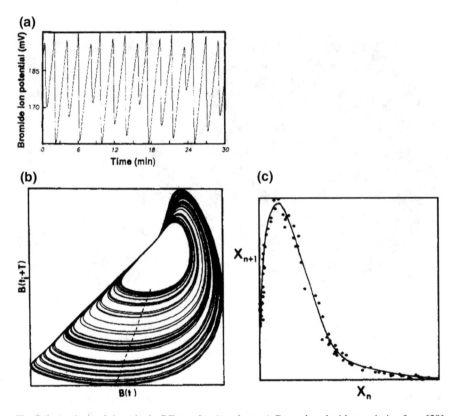

**Fig. 7.6** Analysis of chaos in the BZ reaction (see the text). Reproduced with permission from [29]

a continuously stirred flow reactor and such regimes were found within an interval of the flow rate. The reaction was monitored with the bromide ion specific and with platinum wire electrodes. Similar observations of chaotic oscillations were reported at about the same time by O. Rössler and K. Wegmann [25].

In subsequent experiments, the chaotic nature of observed oscillations was supported by looking at the power spectra of concentration time series: the oscillations exhibited broad-band noise that was much stronger than the instrumental noise found for periodic states [26, 27]. Moreover, the autocorrelation function for concentrations decayed to zero at long times [26]. However, it was still possible that such behavior resulted from the presence of small external perturbations, e.g., in the flow rate. Indeed, the numerical simulations using the reversible Oregonator model did not show the existence of chaotic regimes [28].

In 1983, J. Roux, R. Simoyi and H. L. Swinney [29] (see also the subsequent review[30]) have undertaken a detailed analysis of chaotic dynamics in the BZ reaction. Experimentally, only one signal—the bromide ion potential $B(t)$—was continuously monitored, showing irregular oscillations (Fig. 7.6a). The multidimensional dynamics of the system could however be reconstructed by additionally including the time-shifted signals $B(t+T)$, $B(t+2T)$, $B(t+3T)$... where $T$ is a fixed delay time. As an example, Fig. 7.6b shows the two-dimensional projection of phase trajectories obtained from the time series $B(t+T)$ vs. $B(t)$ with $T = 8.8$ s.

Further analysis was performed by constructing a one-dimensional map. To do this, the authors considered intersection points of the trajectory with the dashed straight line in Fig. 7.6b. The consequent values of the signal $B$ at such points yielded the time series $X_1, X_2, \ldots, X_n, X_{n+1}, \ldots$ If the dependence of $X_{n+1}$ on $X_n$ was plotted (Fig. 7.6c), the points fell on a smooth curve, the one-dimensional map. Thus, even though the behavior was nonperiodic and had a power spectrum with a broad-band noise, the system was nonetheless completely deterministic, i.e. for any $X_n$ the map gave the next value $X_{n+1}$ [30]. The shape of the recurrence map in Fig. 7.6c was a hallmark of the deterministic chaos and, by using it, the Lyapunov exponent (5.26) could be directly determined from the experimental data [29].

Finally, chaotic dynamics could be found in numerical simulations for a mathematical model of the BZ reaction [31]. This model, also based like the Oregonator on the Field–Körös–Noyes scheme [8], had however to consist of 7 differential equations corresponding to 9 reaction steps.

Chemical turbulence (or spatiotemporal chaos) has also been observed in the BZ reaction. In 2000, Qi Ouyang, Harry L. Swinney and Ge Li [32] showed that, under certain conditions, spiral waves in the excitable BZ medium become unstable (Fig. 7.7). Their core begins to meander (Fig. 7.7a), leading to wave breakups in the central part of the wave (Fig. 7.7b). As a result, fragments of new spiral waves are created (Fig. 7.7c). Finally, the entire medium is filled with irregularly moving, colliding and annihilating waves (Fig. 7.7d). This behavior agrees with the theoretical predictions earlier made by M. Bär et al. [33] for the model of a surface chemical reaction (see Chap. 8).

Turing patterns are not observed in the classical BZ system (stationary patterns originally reported in Ref. [17] have turned out to be due to hydrodynamic convection phenomena). This is because the diffusion constants of reacting species in an aqueous solution cannot differ much. The system can however be modified, so that a large difference in diffusion constants of reactants is established. Vladimir Vanag and Irving Epstein investigated [34] in 2001 the BZ reaction in a water-in-oil microemulsion. The microemulsion represented a thermodynamically stable mixture of water, oil, and surfactant, in which the water and surfactant molecules formed nanometer-size droplets filled with water and covered by the surfactant. Such water nanodroplets were dispersed in oil. As the surfactant, sodium bis(2-ethylhehy)sulfosuccinate, or AOT, was used.

The polar BZ reactants and catalyst were confined within the droplets, but nonpolar intermediates ($Br_2$ and $BrO_2$) could leave and enter the oil phase. Because the highly polar species were localized in the droplets, their effective diffusion was due to the Brownian motion of such droplets, their collisions and exchange of material between them. The effective diffusion coefficient determined by random motion of nanodroplets in the oil phase was estimated [34] as varying between $10^{-8}$ and $10^{-7}$ $cm^2/s$. On the other hand, the diffusion coefficient of small molecules dissolved in the oil phase was about $10^{-5}$ $cm^2/s$. Thus, a large difference in the mobilities of

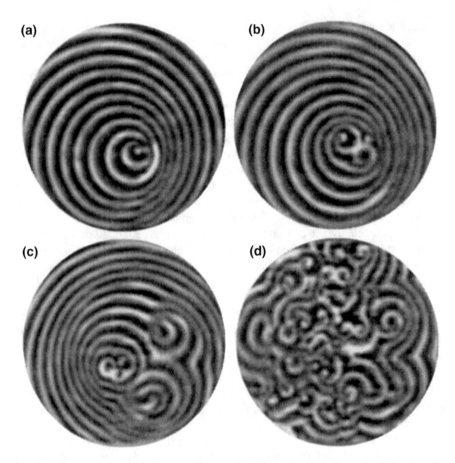

**Fig. 7.7** Development of chemical turbulence in the BZ reaction through the destabilization of a spiral wave. Reproduced with permission from [32]

different species was created. It could be controlled by changing the composition of the system and thus affecting droplet concentrations and the droplet size.

Optical microscopy observations revealed the formation of stationary Turing structures with the characteristic wavelength of 0.18 mm (Fig. 7.8). Labyrinthine (Fig. 7.8a) or hexagonal (Fig. 7.8b) patterns could be seen depending on the volume fraction of the dispersed phase (water plus surfactant).

In his 1952 article, Turing also predicted an instability leading to spontaneous development of traveling waves with a definite frequency and wavelength (this instability is known as the wave bifurcation today). Such an instability is possible only in systems with at least three reacting species and the necessary condition is again that a sufficiently large difference in the mobilities of these species should be present. Due to the symmetry, waves traveling in different directions (but with the same wavelength) begin to grow at the instability point. As a result of the instability, wave

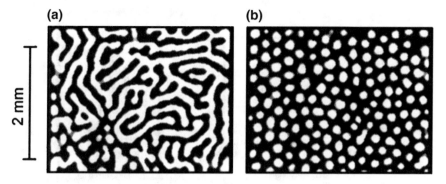

**Fig. 7.8** Turing patterns (**a** labyrinthine and **b** hexagonal) in the AOT water-in-oil microemulsion with the BZ reaction. Reproduced with permission from [34]

patterns representing either propagating or standing waves can set in. A standing wave is a superposition of two waves that propagate in opposite directions.

Standing waves were indeed observed in experiments with BZ microemulsions [34]. They were found for a slightly different composition (i.e., the volume fraction) of the emulsion; their wavelength was 0.23 mm and their period was 91 s. The observed wave patterns were transient: they persisted over several tens of oscillation periods and finally gave way to the uniform steady state.

Finally, synchronization phenomena in large populations of coupled chemical BZ oscillators were also explored. In 2009, A. Taylor et al. [35] performed experiments with a colloid system where porous catalytic particles were suspended in the catalyst-free BZ reaction mixture. The sizes of the particles varied from 50 to 250 μm, with the average of about 100 μm. Typically, about 100,000 such particles were present in the reaction volume; the number density of particles could be varied in the experiments. The reactor was stirred so that the the reactants were well mixed and the system was uniform.

In this system, the reaction took place only within the catalytic particles but intermediate reaction products could be freely transported from one particle to another. The reaction was in the oscillatory regime and thus every catalytic particle effectively represented an individual oscillator. The exchange of reaction products through the solution resulted in interactions—or coupling—between the oscillators. Because of persistent stirring, such coupling was global, i.e. not dependent on spatial positions of the oscillatory microparticles or distances between them.

As shown in Chap. 5, the Kuramoto theory predicts that synchronization in populations of globally coupled oscillators can take place when the strength of interactions between them is increased. The synchronization transition is characterized by an order parameter that yields the collective oscillation amplitude. This order parameter vanishes if oscillations are asynchronous and increases according to Eq. (5.24) above the transition point. For globally coupled electrochemical oscillators, the predicted dependence could be verified in 2002 by I. Kiss, Y. Zhai and J. L. Hudson [36] (Fig. 5.12). This could be however done only for a relatively small system with 64 individual oscillators.

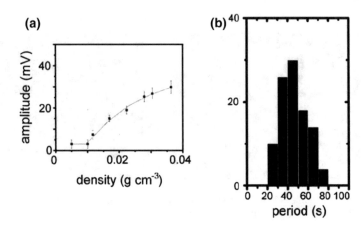

**Fig. 7.9 a** Synchronization transition in a population of globally coupled BZ oscillators.
**b** Distribution of frequencies of individual oscillators. Adapted with permission from [35]

The experiments with the BZ reaction by A. Taylor et al. [35] enabled a test of
the Kuramoto theory for a much larger system with about $10^5$ oscillators. In these
experiments, the strength of coupling could be effectively controlled by changing the
number of colloid particles while keeping the reaction volume fixed. Because of a
dispersion in the sizes of particles, the individual oscillation frequencies varied too,
as illustrated in Fig. 7.9b. The collective oscillation amplitude was characterized by
monitoring the electrochemical potential of the solution wth a Pt electrode sensitive
to a change in the concentration of $HBrO_2$ relative to $Br^-$. Figure 7.9a shows the
dependence of the collective oscillation amplitude on the population density of catalytic particles that determines the coupling strength. The synchronization transition
is clearly seen.

The Belousov–Zhabotinsky reaction, and particularly its photosensitive modification, have been moreover broadly used in the studies on control of nonequilibrium
pattern formation and self-organization in chemical systems. We return to discussion
of such experiments in Chap. 10.

## References

1. A.P. Belousov, The periodically acting chemical reaction and its mechanism, unpublished
   (1951). Reproduced from his archive in *Avtovolnovye Protsessy v Sistemakh s Diffuziey*
   (Autowave Processes in Systems with Diffusion), ed. by M.T. Grekhova et al. (Nizhni Nov-
   gorod, Institute of Applied Physics of the USSR Academy of Sciences, 1981), pp. 176–189;
   English translation in *Oscillations and Traveling Waves in Chemical Systems*, ed. by R. Field,
   M. Burger (Wiley, New York, 1985) pp. 605–613
2. B.P. Belousov, Periodically acting reaction and its mechanism, in *Sbornik Referatov po Radi-
   atsionnoy Medicine* (Collection of Reports on Radiation Medicine) (Biophysics Institute of the
   Ministry of Health of the USSR, 1958), pp. 145–147

3. A. Winfree, J. Chem. Educ. **61**, 661 (1984)
4. A.M. Zhabotinsky, Biofizika **9**, 306 (1964)
5. A.M. Zhabotinsky, Dokl. Akad. Nauk SSR **157**, 392 (1964)
6. A.M. Zhabotinsky, Chaos **1**, 379 (1991)
7. H. Degn, Nature **213**, 589 (1967)
8. R.J. Field, E. Körös, R.M. Noyes, J. Am. Chem. Soc. **94**, 8649 (1972)
9. L. Györgyi, T. Turanyi, R.J. Field, J. Phys. Chem. **94**, 7162 (1990)
10. R.J. Field, R.M. Noyes, J. Phys. Chem. **60**, 1877 (1974)
11. J.J. Tyson, P.C. Fife, J. Chem. Phys. **73**, 2224 (1980)
12. V.A. Vavilin, A.M. Zhabotinsky, A.N. Zaikin, Russ. J. Phys. Chem. **42**, 3091 (1968)
13. A.N. Zaikin, A.M. Zhabotinsky, in *Biological and Biochemical Oscillators*, ed. by B. Chance et al. (Academic Press, New York, 1973) p. 81
14. A. Wiener, A. Rosenblueth, Arch. Inst. Cardiol. Mex. **16**, 205 (1946)
15. A. Wiener, A. Rosenblueth, in *Kiberneticheskiy Sbornik*, vol. 3, ed. by A.A. Lyapunov et al. (Foreign Literature Publishing House, Moscow, 1961), pp. 7–56
16. A.N. Zaikin, A.M. Zhabotinsky, Nature **225**, 535 (1970)
17. A.M. Zhabotinsky, A.N. Zaikin, J. Theor. Biol. **40**, 45 (1973)
18. C. Vidal, A. Pagola, J. Phys. Chem. **93**, 2711 (1989)
19. A.M. Zhabotinsky, *Investigations of Homogeneous Chemical Auto-Oscillating Systems* (Institute of Biological Physics of the USSR Academy of Sciences, Puschino, 1970)
20. A.M. Zhabotinsky, A.N. Zaikin, in *Oscillatory Processes in Biological and Chemical Systems*, vol. 2, ed. by E.E. Sel'kov (Institute of Biological Physics of the USSR Academy of Sciences, Puschino, 1971), p. 273
21. A.T. Winfree, Science **175**, 634 (1972)
22. A.T. Winfree, Science **181**, 937 (1973)
23. D. Stock, S.C. Müller, Phys. D **96**, 396 (1996)
24. R.A. Schmitz, K.R. Graziani, J.L. Hudson, J. Chem. Phys. **67**, 3040 (1977)
25. O.E. Rössler, K. Wegmann, Nature **271**, 89 (1978)
26. C. Vidal, J.C. Roux, S. Bachelart, A. Rossi, Ann. N.Y. Acad. Sci. **357**, 377 (1980)
27. C. Vidal, S. Bachelart, A. Rossi, J. Phys. **43**, 7 (1982)
28. K. Showalter, R.M. Noyes, K. Bar-Eli, J. Chem. Phys. **69**, 2514 (1978)
29. J.C. Roux, R.H. Simoyi, H.L. Swinney, Phys. D **8**, 257 (1983)
30. F. Argoul, A. Arneodo, P. Richetti, J.C. Roux, H.L. Swinney, Acc. Chem. Res. **20**, 436 (1987)
31. P. Richetti, J.C. Roux, F. Argoul, A. Arneodo, J. Chem. Phys. **86**, 3339 (1989)
32. Q. Ouyang, H.L. Swinney, G. Li, Phys. Rev. Lett. **84**, 1047 (2000)
33. M. Bär, M. Hildebrand, M. Eiswirth, M. Falcke, H. Engel, M. Neufeld, Chaos **4**, 499 (1994)
34. V.K. Vanag, I.R. Epstein, Phys. Rev. Lett. **87**, 228301 (2001)
35. A.F. Taylor, M.R. Tinsley, F. Wang, Z. Huang, K. Showalter, Science **323**, 614 (2009)
36. I. Kiss, Y. Zhai, J.L. Hudson, Science **296**, 1676 (2002)

# Chapter 8
# Catalytic Reactions at Solid Surfaces

Chemical reactions at solid surfaces form the basis of heterogeneous catalysis. Among them, the oxidation of CO with $O_2$ to $CO_2$ is the simplest one. It also served widely as a system for studying kinetic oscillations and the phenomena of spatio-temporal pattern formation. The first report on (isothermal) oscillations in this reaction can be found in a short communication by Peter Hugo [1] that has appeared in 1969. Later, he studied this effect in more detail [2], as this was also done in the same laboratory by E. Wicke et al. [3]. Figure 8.1 gives an example for periodic variations of the CO and $CO_2$ concentrations with time [2]. Although no explanation could be given, it was speculated that CO might be adsorbed in two different states which differ in reactivity. As will be shown below, the surface can indeed switch between two states of reactivity and this accounts for the observed effects.

Meanwhile, oscillatory kinetics has been reported for many heterogeneously catalyzed reactions. The experiments were undertaken with 'real' (supported) catalysts under ambient conditions [4], as well as with single crystal surfaces under ultrahigh vacuum (UHV) conditions serving as model systems [5]. The latter experiments provide the most detailed insights into the underlying mechanisms and the general phenomenology of spatiotemporal pattern formation. The UHV studies are performed with bulk samples under the conditions of extremely low pressure where temperature changes associated with varying reaction rates are negligible and possible concentration gradients in the gas phase are practically instantaneously levelled out.

Kinetic oscillations with such a system were observed for the first time in 1981 for the catalytic oxidation of CO at a polycrystalline platinum foil. Figure 8.2 shows the original data in which the rate of $CO_2$ formation, together with the partial pressures of CO and $O_2$ and the variation of the work function, are plotted as functions of time [6]. The latter quantity reflects the state of adsorption and clearly indicates that the effect is connected with a variation of the surface properties. Subsequent experiments with clean single crystal surfaces revealed that this effect is absent with the most densely packed Pt(111) surface, but present within a certain range of parameters with the Pt(100) surface the structure of which may undergo transformation under the influence of CO adsorption [7]. Even more pronounced effects were later found

© Springer International Publishing AG 2017                                    105
A.S. Mikhailov and G. Ertl, *Chemical Complexity*, The Frontiers Collection,
DOI 10.1007/978-3-319-57377-9_8

**Fig. 8.1** Temporal
oscillations in the rate of
$CO_2$ production and CO
consumption in CO
oxidation at a single Pt
catalyst particle. Reproduced
from [2]

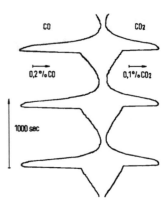

with the Pt(110) surface which presents the richest variety of phenomena and will
serve as the main system for the rest of this chapter.

Figure 8.3 shows how in this case under steady-state flow conditions the rate of
$CO_2$ formation varies with time if the temperature and the CO partial pressure are
kept fixed and at the point marked by an arrow the $O_2$ partial pressure is increased
stepwise from 2.0 to $2.7 \times 10^{-4}$ mbar. The rate increases slowly and then becomes
oscillatory with the constant amplitude. These oscillations occur within a narrow
range of external parameters where the structure of the Pt(110) surface alternates
periodically between the reconstructed $1 \times 2$ and the normal $1 \times 1$ phases.

These two modifications of the surface structure are depicted in Fig. 8.4. The clean
surface exhibits a reconstructed 'missing row' $1 \times 2$-structure which is energetically
more stable than the bulk-like $1 \times 1$-structure. For the latter, however, the energy
of CO adsorption is higher so that it is formed as soon as the surface concentration
or coverage of CO exceeds a certain critical value. At 300 K this transformation is
initiated by homogeneous nucleation of $1 \times 1$-patches exposed from the second layer.
At elevated temperatures lateral displacements of longer atomic chains take place,
and the enhanced surface mobility enables the formation of larger $1 \times 1$-islands [8].

This transformation is driven by the difference in CO adsorption energy in both
phases. On the other hand, the probability of dissociative adsorption (characterized
by the sticking coefficient $s_{O_2}$) of $O_2$ is higher on the $1 \times 1$-than the $1 \times 2$-phase.
Since the oscillations are occurring under the conditions for which oxygen adsorption
is rate-limiting, their origin may now be readily rationalized: Let us start with the
$1 \times 2$-surface and a ratio of the partial pressures of the reactants that is such that
enough CO is adsorbed to cause the transformation into the $1 \times 1$-phase. Under these
conditions, $O_2$ will be adsorbed and the $O_{ad}$ atoms will react with $CO_{ad}$ more readily,
so that the coverage of the latter decreases below the critical value for the stabilization
of the $1 \times 1$-structure. Now a transformation back to the $1 \times 2$-phase takes place at
which the oxygen sticking coefficient is smaller, so that the CO coverage increases
again and the cycle will be repeated.

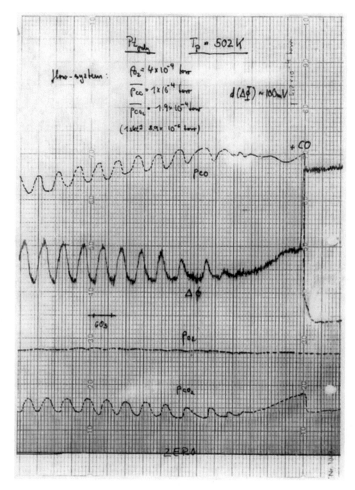

**Fig. 8.2** The original strip chart showing the onset of kinetic oscillations in the catalytic oxidation of CO on a polycrystalline Pt foil. The time is going from *right* to *left* with 60 s marked. J. Rüstig and G. Ertl, unpublished (1981)

Theoretical description of this mechanism can be performed within a mean-field model. This kinetic model uses three variables $u$, $v$ and $w$. The first two of them represent surface coverages of CO and O, respectively. The surface coverage $\theta$ for a chemical species yields the fraction of adsorption sites for this species that are actually occupied by it; it varies therefore between zero and one. Thus, we have $u = \theta_{CO}$ and $v = \theta_O$. The variable $w$ specifies the fraction of the surface existing in the $1 \times 1$-phase, i.e. $w = \theta_{1 \times 1}$. Note that then $\theta_{1 \times 2} = 1 - w$.

The reaction consists of four elementary steps:

**Fig. 8.3** Onset of kinetic oscillations in the oxidation of CO on a Pt(110) surface. $T = 470$ K, $p_{CO} = 2.0 \rightarrow 2.7 \times 10^{-4}$ mbar. Reproduced from [9]

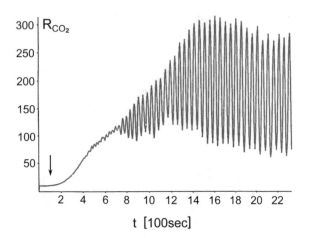

**Fig. 8.4** The $1 \times 2$–$1 \times 1$ structural transformation of the Pt(110) surface

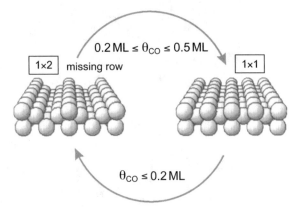

$$CO + * \rightleftarrows CO_{ad} \qquad (8.1)$$

$$O_2 + 2* \rightarrow 2O_{ad} \qquad (8.2)$$

$$CO_{ad} + O_{ad} \rightarrow CO_2 + 2* \qquad (8.3)$$

$$Pt(1 \times 2) \xrightarrow{CO} Pt(1 \times 1). \qquad (8.4)$$

where the symbol $*$ denotes a vacant surface site.

The step (8.1) represents CO adsorption and desorption. The adsorption rate is given by $s_{CO}p_{CO}$ where $p_{CO}$ is the CO partial pressure in the gas phase. The process of CO adsorption involves the formation of a weakly bound surface state and hence it is sensitive to the local presence of several vacant sites. Therefore, the sticking coefficient $s_{CO}$ is not simply proportional to the fraction $1 - u$ of vacant states. Instead, it has a more complex dependence $s_{CO} = k_1(1 - u^3)$ where $k_1$ is the hitting rate constant. Note that surface oxygen cannot block the adsorption of CO and therefore the sticking coefficient is independent of $v$. The desorption of CO from the surface has the rate $k_2$.

The second step (8.2) is the dissociative adsorption of oxygen, it is characterized by the rate $s_{O_2} p_{O_2}$ where $p_{O_2}$ is the partial pressure of oxygen. The sticking coefficient $s_{O_2}$ of oxygen depends on the fraction of available surface sites and also on the structural state of the surface characterized by the variable $w$. Both oxygen ($O_{ad}$) and carbon monoxide ($CO_{ad}$) atoms on the surface block the dissociative adsorption of $O_2$. Therefore, its adsorption rate is proportional to $(1 - u - v)^2$. Thus, the coefficient $s_{O_2}$ for oxygen is $k_4 [s_1 w + s_2(1 - w)](1 - u - v)^2$ where $k_4$ is the hitting rate constant for oxygen molecules. The sticking coefficient is enhanced in the $1 \times 1$ phase and therefore $s_1 > s_2$. Oxygen is strongly bound on the Pt surface and its desorption is negligible.

The third step is the oxidation reaction between the adsorbed species. Its rate is $k_3 u v$ where $k_3$ is the respective rate constant. Note that the produced $CO_2$ immediately leaves the surface and the reverse reaction cannot occur.

Finally, the step (8.4) schematically denotes structural surface changes that are induced by the adsorbed carbon monoxide. The actual mechanism of surface transformation is complex and such effects are only phenomenologically taken into account in the considered mechanism. The rate of change of the variable $w$ is given by $k_5[f(u) - w]$ where $k_5$ is a rate constant and $f(u)$ is a steplike function of the CO coverage $u$ that changes from zero to one when the coverage $u$ exceeds a threshold.

Thus, kinetic equations of the model have the form [10]

$$\frac{du}{dt} = k_1 p_{CO}(1 - u^3) - k_2 u - k_3 u v \tag{8.5}$$

$$\frac{dv}{dt} = k_4 p_{O_2} [s_1 w + s_2(1 - w)] (1 - u - v)^2 - k_3 u v \tag{8.6}$$

$$\frac{dw}{dt} = k_5[f(u) - w]. \tag{8.7}$$

It is important that the structural surface transition is slow as compared to the reaction and adsorption processes: the respective time scales differ by several orders of magnitude. Therefore, this phase transition can act as a switch.

In absence of the phase transition, i.e. if $w = const$, the first two equations of the model yield bistability. One of the states is only weakly reactive and corresponds to "poisoning" of the surface by carbon monoxide. In this state, almost all sites are occupied by CO and there are only few vacant sites where oxygen can arrive. The other stationary state is highly reactive, with many vacant sites persistently created in reaction events.

The oscillations develop as a result of an interplay between the bistability and the structural change in the catalyst. As the highly reactive state is approached, the CO coverage decreases and the surface gets therefore changed into the $1 \times 2$ structure.

**Fig. 8.5** Oscillations in the rate of CO oxidation reaction on Pt(110), in the CO and O coverages $u$ and $v$, and in the fraction $w$ of the surface in the $1 \times 1$-phase. Simulations of the kinetic model (8.5)–(8.7) for the parameters $T = 540$ K, $p_{O_2} = 6.7 \times 10^{-5}$ mbar, $p_{CO} = 3 \times 10^{-5}$ mbar. Reproduced from [10]

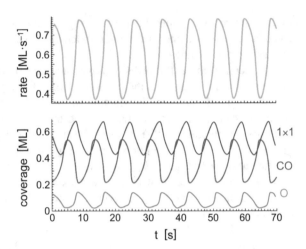

In this state, the adsorption rate of oxygen is however lower, so that the system becomes to return to the CO-poisoned state. When the CO coverage is higher, the surface returns back to the $1 \times 1$ state, more oxygen begins to arrive, and the cycle is repeated again.

Figure 8.5 shows the results of simulations for a typical set of parameters. Note that the O and CO coverages are strictly anti-correlated, while $w$ follows them with a certain phase shift. The oscillation period is about 5 s; it is mostly determined by the long characteristic time of the surface structure switch.

Although different chemical processes are involved, the mathematical description (8.5)–(8.7) resembles the Oregonator model for the BZ reaction. A catalyst is alternating between the low- and high-activity states and, because of this, oscillations take place. In addition to oscillations, the model (8.5)–(8.7) can exhibit bistability or excitable dynamics. Moreover, chaotic oscillations can also occur and they were indeed experimentally observed.

Figure 8.6 gives an example of the transition to chaos with the CO partial pressure used as the control parameter. Here, the upper row shows the experimental time series for the reaction rate. The lower row exhibits the corresponding phase portraits constructed from the time series by the time delay method [11], similar to Fig. 7.7b for the oscillatory BZ reaction. Initially, simple periodic oscillations are observed (Fig. 8.6a). If the CO partial pressure is then slightly lowered, a bifurcation takes place and the period becomes doubled, so that the oscillations are characterized by alternating small and larger amplitudes (Fig. 8.6b). The next period doubling occurs upon a slight lowering of $p_{CO}$ (Fig. 8.6c). A very small further decrease of $p_{CO}$ causes however a qualitative change: the time series becomes chaotic (Fig. 8.6d).

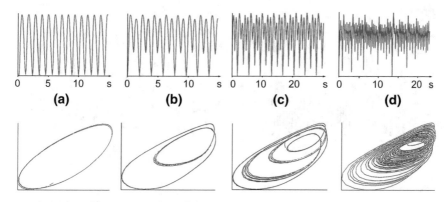

**Fig. 8.6** Transition to chaos via a sequence of period doublings in the CO oxidation on Pt(110). The *upper row* shows the experimental time series, while the *lower row* exhibits the corresponding phase portraits. The CO partial pressure $p_{CO}$ is **a** $1.65 \times 10^{-4}$ mbar, **b** $1.62 \times 10^{-4}$ mbar, **c** $1.60 \times 10^{-4}$ mbar, and **d** $1.56 \times 10^{-4}$ mbar. Other parameters are $T = 550$ K, $p_{O_2} = 4.0 \times 10^{-4}$ mbar. Reproduced from [12]

These experimental results agree well with the theoretical scenario of the transition to chaos through a series of period doubling bifurcations illustrated in Fig. 5.15 for the Rössler model.

Novel effects come into play if, in addition to the oscillatory kinetics under constant external parameters, one of these parameters is periodically varied so that forced oscillations result. As noted in Chap. 5, external forcing can lead to entrainment effects. The experimental data for external forcing of oscillations in the CO + O system on Pt(110) is reproduced in Fig. 8.7. Here the $O_2$ partial pressure is periodically modulated and the change of the work function serves to monitor the variation of the reaction rate. The examples show subharmonic 1:2 entrainment (Fig. 8.7a), superharmonic 2:1 (Fig. 8.7b) and 7:2 (Fig. 8.7c) entrainments and the non-entrained quasi-periodic regime (Fig. 8.7d).

The entrainment windows, or Arnold tongues, for different experimentally observed syncronization regimes are shown in the upper part of Fig. 8.8. Here the shaded area corresponds to the non-entrained regime of quasi-periodic oscillations. The largest window is found for the harmonic 1:1 resonance, but the 2:1 window is also relatively large. For comparison, the lower part in Fig. 8.8 shows theoretical results based on numerical simulations of the model (8.5)–(8.7). The boundaries of different entrainment windows correspond to different kinds of bifurcations in this dynamical system.

To model observed oscillations, Eqs. (8.5)–(8.7) were used. These equations however tacitly assume that the concentrations of reacting species are strongly synchronized over the surface and hence laterally uniform. Otherwise, superposition of uncorrelated contributions from different regions would have caused averaging out and disappearance of the oscillations. For homogeneous reactions in solutions (such as the BZ reaction), spatial uniformity results from continuous stirring. But there is no stirring in surface reactions and synchronization cannot be a consequence of it.

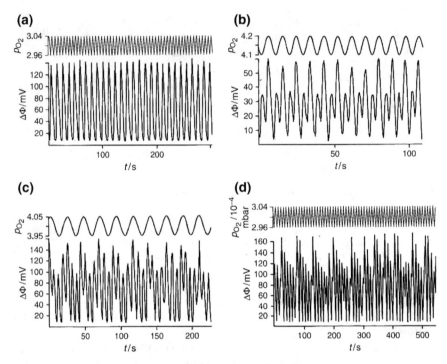

**Fig. 8.7** Experimental data of time series for periodically forced oscillations in the CO oxidation on a Pt(110) surface: **a** subharmonic 1:2 entrainment, **b** superharmonic 2:1 entrainment, **c** superharmonic 7:2 entrainment, and **d** non-entrained quasi-periodic oscillations. Variation in the work function, characterizing the reaction rate, is displayed. The corresponding forcing signal $p_{O_2}(t)$ is shown *above each plot*. Reproduced from [13]

An effective counterpart to rapid stirring in solution can be present in the experiments at very low pressures. Under ultrahigh vacuum (UHV) conditions, the mean free path of molecules in the gas typically exceeds the dimension of the reactor. Their mean velocity is so large (approximately $10^3$ m/s) that concentration gradients in the gas phase due to varying surface concentrations practically disappear instantaneously. Therefore, partial pressures of gaseous reactants are the same for all elements of the surface and, moreover, they are collectively determined by the processes in all of them. Instantaneous interactions through the gas phase lead to "global coupling" in surface reactions at UHV conditions.

This effect of global coupling may be enough to synchronize even different samples in a reactor vessel [15]. The just described experiments with external forcing demonstrate that variations of the partial pressures far below 1% can lead to significant synchronisation effects. In Chap. 10, we will show that global coupling through the gas phase can be also used to control chaos in surface reactions.

Another counterpart to stirring is possible if changes in the reaction enthalpy accompanying local variations of surface concentrations of reactive species give rise

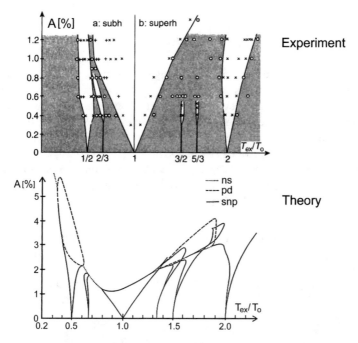

**Fig. 8.8** Experimental (*above*) and theoretical (*below*) entrainment diagrams for forced oscillations in the catalytic CO oxidation on a Pt(110) surface. The ratio of the forcing period $T_{ex}$ to the natural oscillation period $T_0$ is plotted along the *horizontal axes*. The forcing amplitude increases in the vertical direction. In the experimental diagram, the *shaded areas* indicate the non-entrained quasi-periodic regimes. In the theoretical diagram, the *lines* show the boundaries of different bifurcations: *ns* Neimark–Sacker, *pd* period-doubling, *snp* saddle-node. Reproduced from [14]

to lateral temperature gradients. Coupling between different elements of a catalytic surface takes then place through the heat flux. This mechanism will generally prevail at the higher pressures and also with supported catalysts where temperature variations up to 100 K over the surface can arise. The characteristic "diffusion length" (i.e., the heat conduction length) is in this case of the order of 1 mm. Thus, even if the catalyst consists of separate particles, this length can be much larger than the mean separation between the particles or the diameters of their individual crystal planes. As a consequence, such a system can again be regarded as being uniform on its relevant length scale. However, the observed phenomena will be then rather dominated by heat effects than by the surface chemistry details.

Surface diffusion of adsorbed species always takes place and dominates over heat conduction effects in the studies with bulk single crystals under low-pressure (UHV) conditions where lateral temperature gradients are negligible. In this case, characteristic length scales will be typically of the order of $10^{-5}$ cm, so that suitable experimental techniques are required to image such truly two-dimensional patterns. Mathematical modelling of these phenomena must include diffusion of adsorbed species, so that the resulting equations will generally be of the type

$$\frac{\partial c_i}{\partial t} = f_i\left(c_j, p_k\right) + D_i \nabla^2 c_i \tag{8.8}$$

where $c_i$ are surface concentrations (coverages) of various reacting species $i$, $p_k$ are partial pressures of gaseous components, and $D_i$ are surface diffusion constants. Additionally, equations for the evolution of partial pressures can be included to take into account the above mentioned global coupling effects. Moreover, the equations describing structural changes have to be added if such changes take place.

In our preferred example of CO oxidation on a Pt(110) surface, adsorbed oxygen is practically immobile under typical experimental conditions and therefore only the surface diffusion of adsorbed CO needs to be considered. Thus, the mathematical model of this reaction takes the form

$$\frac{\partial u}{\partial t} = k_1 p_{CO}(1 - u^3) - k_2 u - k_3 u v + D \nabla^2 u \tag{8.9}$$

$$\frac{\partial v}{\partial t} = k_4 p_{O_2} \left[s_1 w + s_2(1 - w)\right] (1 - u - v)^2 - k_3 u v \tag{8.10}$$

$$\frac{\partial w}{\partial t} = k_5 [f(u) - w] \tag{8.11}$$

where $D$ is the CO surface diffusion constant. The adsorption rates are proportional to partial pressures $p_{CO}$ and $p_{O_2}$ of CO and oxygen in the gas phase. The model can be complemented by the equations for these partial pressures if their significant variation during an experiment takes place. The CO diffusion on Pt(110) is anisotropic and this can also be taken into account by introducing appropriate anisotropy terms.

Imaging of the concentration patterns is most conveniently performed by means of photoemission electron microscopy (PEEM) [16]. The principle of this technique is based on the differing dipole moments of the adsorbed species that give rise to variation of the local electron work function, which in turn affects the yield of electrons emitted from a surface upon irradiation with ultraviolet light. The lateral intensity distribution of these photoelectrons reflects the kind and concentration of adsorbed species and is imaged through a system of electrostatic lenses onto a channel plate and a fluorescent screen. From there, the image can be recorded using a CCD camera, and typical resolutions of $2 \times 10^{-5}$ cm and 20 ms respectively can be achieved. Since the adsorbed oxygen causes a stronger increase of the work function than the adsorbed CO, areas predominantly covered by oxygen appear as dark, while those mainly covered by CO are gray in the PEEM images.

By employing the PEEM system, a wealth of time-dependent lateral concentration patterns was observed depending on the chosen external parameters, i.e. on the temperature and CO and $O_2$ partial pressures.

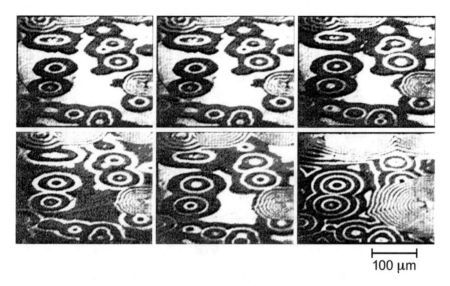

**Fig. 8.9** Target patterns in the CO oxidation on a Pt(110) surface. A series of images recorded at intervals of 4.1 s by photoemission electron microscopy (PEEM). The parameters are $T = 427$ K, $p_{O_2} = 4.0 \times 10^{-4}$ mbar, $p_{CO} = 3 \times 10^{-5}$ mbar. Reproduced from [18]

As the first example, Fig. 8.9 shows typical "target" patterns, already known from the BZ reaction. Now, they are however elliptical instead of having a circular shape. This demonstrates directly the role of surface diffusion of adsorbed CO which is faster along the atomic rows of the surface structure shown in Fig. 8.4. The expansion and propagation of the elliptical patterns is superseded by a periodic variation of the brightness of the background signaling the participation of global coupling through the gas phase. Theoretical modeling could in this case be achieved by assuming defects acting as the pacemakers [17].

In addition to the oscillatory regime, the excitable conditions and the situation of double metastability can be found for CO oxidation on Pt(110). A schematic diagram, indicating regions in the parameter plane $(p_{CO}, T)$ where different kinds of patterns were observed [19], is shown in Fig. 8.10.

At low partial pressures of CO, the surface is in the uniform oxygen covered state. Increasing the CO pressure leads from this stable state to the double metastability region. Further rise of the CO pressure leads to excitability and the growth of rotating oxygen spiral waves. At even higher CO pressure, a narrow range of low excitability is reached where spirals disappear in favor of traveling solitary wave fragments. In the limit of very high CO pressure, the surface is in the uniform CO covered state.

Generally, a bistable system can be in two states which are both stable against small perturbations. One of these states is stable and the other is metastable. If a sufficiently strong local perturbation is applied, a nucleus of the stable state becomes created. This nucleus grows and its moving boundary represents a propagating transition front. After its spreading, the medium is transferred to the stable uniform state.

For CO oxidation on Pt(110), the surface can be in the states that are predominantly covered either by O or CO. However, the simple bistability which we have just

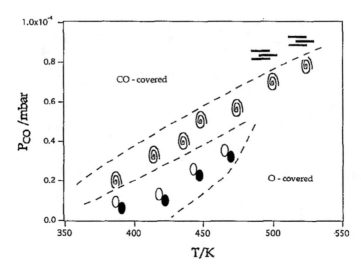

**Fig. 8.10** Experimental diagram illustrating the existence regions of different wave patterns. Elliptic spreading O and CO fronts are characteristic for the double metastability regime. Spirals are formed in the excitability region. At the limit of low excitability (high $p_{CO}$ and high $T$), the intrinsic core of the spiral diverges and flat solitary waves may be formed as symbolized by the *parallel bars*. The oxygen partial pressure is $p_{O_2} = 4 \times 10^{-4}$ mbar. Reproduced from [19]

described is not observed. Instead, *both* uniform states turn out to be metastable and nucleation of the opposite phase through strong local perturbations (i.e., the surface defects) takes place. Therefore, for the same set of control parameters, both O and CO fronts are observed. In Chap. 6, we have noted that such complex bistability (or *double metastability*) is possible for two-component activator-inhibitor systems. This phenomenon is common for CO oxidation on Pt(110) and observed in a broad region in the parameter plane (Fig. 8.10).

Above the double metastability region (i.e., at the higher CO pressure), the excitable regime sets on. Here, predominantly oxygen covered pulses can propagate on the background of the CO covered state. It should be noted that, under experimental conditions, it is difficult to draw an exact border between the two regimes. As shown in Chap. 6, excitation waves can be already observed inside the double metastability region where they represent bound states of two opposite traveling fronts.

The excitation waves often develop into spirals [19]. Such a situation is shown in Fig. 8.11. It is remarkable that under identical external conditions not all spirals exhibit the same periods and wavelengths. This has to be attributed to pinning of the spiral cores to surface defects of the varying size and kinetic properties. Extended defects may even form the cores of multi-armed spirals. The waves propagate with a velocity of several μm/s. On the atomic scale, about 10 $CO_2$ molecules per second and site are produced within the dark (O-covered) regions, demonstrating the quite different time scales of microscopic and mesoscopic processes.

While most of the observed spirals were pinned by the defects, freely rotating spiral waves could also be observed at the upper boundary of the excitable region

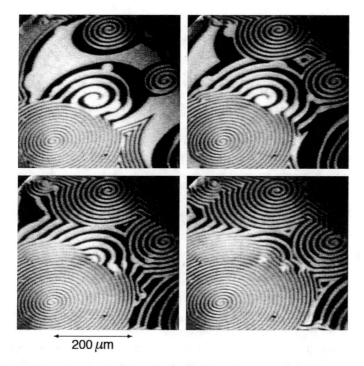

**Fig. 8.11** Rotating spiral waves on a Pt(110) surface during the CO oxidation reaction. A sequence of PEEM images taken at intervals of 30 s. The parameters are $T = 448$ K, $p_{O_2} = 4 \times 10^{-4}$ mbar, $p_{CO} = 4.3 \times 10^{-5}$ mbar. Reproduced from [19]

where the excitability is decreased. The existence of free spirals was demonstrated in the experiments with external forcing described below in Chap. 10. As the excitability was further decreased by raising the CO partial pressure, excitation waves stopped to coil into spirals. Instead, solitary flat traveling fragments of excitation waves were observed. Their propagation direction was determined by the surface anisotropy.

Theoretical modeling of wave propagation phenomena in CO oxidation on Pt(110) was performed [20] by using Eqs. (8.9)–(8.11). The analysis of the model revealed that it predicts all principal observed regimes, i.e. oscillations, double metastability and the excitable behavior. The theoretically constructed diagram for such regimes in the parameter plane $(p_{CO}, T)$ agreed qualitatively with the experimental results [20].

It was also further shown that, by adiabatic elimination of the oxygen coverage variable $v$, a two-component model with the variables $u$ and $w$ can be derived. Remarkably, this reduced description is similar to the FitzHugh–Nagumo model (6.10)–(6.11) which is used to describe wave propagation along the nerves. The adsorbed CO plays the role of an activator species, whereas the structural surface phase transition provides an inhibition mechanism.

It was shown that the model reproduces the effects of double metastability (and this term was also coined out in the same study). When an O front was running at a higher velocity into the preceding CO front, the formation of a stable traveling

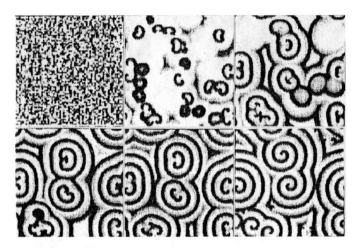

**Fig. 8.12** Numerical simulation of spiral waves on a Pt(110) surface during CO oxidation. Reproduced from [21]

CO pulse that represented the bound state of the fronts and effectively an excitation wave, was observed in numerical simulations.

Simulations of spiral waves in CO oxidation on Pt(110) were also performed. An example of such a simulation is reproduced in Fig. 8.12. Starting from random distribution of the adsorbates, nucleation of spiral waves developing into propagating waves occurs. Because structural defects were not present in this simulation, all spirals are free. Therefore, their rotation periods are the same and they coexist. Under certain conditions, spiral cores (unless pinned to defects) may meander across the surface. If the parameters were slightly varied, the simulations revealed the break-up of the spirals and a transition to chemical turbulence, as shown in Fig. 8.13.

Similar irregular patterns of spiral waves were also experimentally observed. A PEEM image illustrating this situation is shown in Fig. 8.14. Chemical turbulence was furthermore seen in the experiments in the oscillatory regime. It could be then efficiently controlled by introducing an artificial global feedback as discussed further in Chap. 10.

The reaction of CO oxidation on Pt(110) provides the most rich variety of oscillatory and spatio-temporal phenomena. Analogous behavior has however been also observed in a number of different surface reactions too. For example, excitability was found in the reduction of NO with hydrogen on a Rh(110) surface. This surface dissociates NO into $N_{ad}$ and $O_{ad}$ that react with $H_{ad}$ to form $N_2$ and $H_2O$. In PEEM images, oxygen corresponds to the dark area, whereas the nitrogen-covered and bare surface appear as bright. There are two types of the adsorbate-induced surface reconstruction: oxygen generates a missing-row reconstruction consisting of atomic troughs along the [110] direction, whereas nitrogen gives rise to Rh-N chains oriented perpendicular to the direction of oxygen-induced troughs. Under UHV conditions, target patterns and spiral waves could be seen.

**Fig. 8.13**  Development of chemical turbulence through repeated break-ups of spiral waves in the model of CO oxidation on Pt(110). Reproduced from [22]

**Fig. 8.14**  A PEEM image snapshot from the observation showing break-ups of spiral waves and development of spatio-temporal chaos during the CO oxidation on a Pt(110) surface. Reproduced from [18]

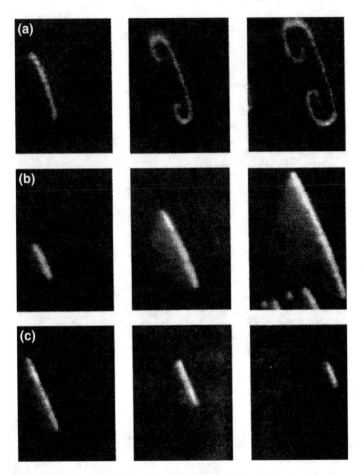

**Fig. 8.15** Traveling fragments under different experimental conditions. **a** Development of two spiral waves ($p_{H_2} = 6.6 \times 10^{-6}$ mbar, $T = 580$ K), **b** a steadily traveling and expanding fragment ($p_{H_2} = 4.6 \times 10^{-6}$ mbar, $T = 620$ K), and **c** a shrinking fragment ($p_{H_2} = 1.8 \times 10^{-6}$ mbar, $T = 620$ K); $p_{NO} = 1.8 \times 10^{-6}$ mbar. The frame size is $75 \times 100\,\mu$m, the interval between the snapshots is **a** 3 s and **b**, **c** 30 s. Subsequent frames are shifted to accommodate the entire traveling object. Reproduced from [23]

Additionally, flat fragments of excitation waves were observed [23]. They travelled at a constant velocity of 1.28 μm/s along the [001] direction while expanding in the orthogonal direction (Fig. 8.15b). The trajectories of their ends represented straight lines; the angular width of the cone was the same for all of them. When fragments collided, annihilation occurred. These structures were sensitive to pressure variations. When $p_{H_2}$ was increased, the initial flat fragment gave rise to a pair of spiral waves (Fig. 8.15a). If $p_{H_2}$ was decreased, the fragments began to shrink (Fig. 8.15c).

Such wave behavior can be understood [23] in the framework of the Wiener–Rosenblueth model for broken waves (see Chap. 6) for anisotropic excitable media. In this model, each small element of the wave moves in its normal direction with the velocity $V = V_0 - DK$ where $K$ is the local wave curvature and $D$ is a coefficient. If the wave is broken and has a free end, it grows or contracts tangentially at this end with the velocity $G = G_0 - \gamma K_0$ where $K_0$ is the curvature at the end. The sign of $G_0$ is important: if $G_0 > 0$, the wave grows and coils at its free end, so that a spiral becomes formed (Figs. 6.7c,d, 6.8 in Chap. 6). If $G_0$ is negative, the wave shrinks (Fig. 6.7a).

If the medium is anisotropic, all velocities can additionally depend on the propagation direction of a wave element. For simplicity, however, we assume that only the growth velocity depends on it, i.e. $G_0 = G_0(\alpha)$ where $\alpha$ is the propagation angle. What behavior should be found if $G_0$ changes its sign and becomes negative as the angle is increased? The initial flat segment will begin to grow and coil at its two open ends so that the pair of spirals would tend to be formed. But, as the wave coils, the propagation angle of the end elements will increase and the growth velocity will become smaller. When the angle $\alpha_0$ is reached, such that $G_0(\alpha_0) = 0$, the growth at the free end will not any longer take place. Hence, the end points will move along straight lines. As a result, a steadily traveling and expanding wave fragment will be produced, as shown in Fig. 8.16b. The angular width of the expansion cone will be given by $\alpha_0$. When the parameters change, this angle can increase and, eventually, a situation can be reached when $G_0$ is positive for all angles. In this case, the initial flat fragment gives rise to a pair of spiral waves (Fig. 8.16a). On the other hand, if $G_0$ is negative already for $\alpha = 0$, the flat wave fragment will shrink (Fig. 8.16c).

The same effects could be seen [23] in direct simulations for an anisotropic version of the FitzHugh–Nagumo model (6.10)–(6.11). Here, it should be noted that a simple anisotropy, such that the diffusion coefficient of the mobile activator species $u$ takes two constant values $D_x$ and $D_y$ along the coordinate axes $x$ and $y$, is not sufficient to account for it. Indeed, this anisotropy can be eliminated by an appropriate rescaling of the coordinates. However, the anisotropy can depend on the structural state of the

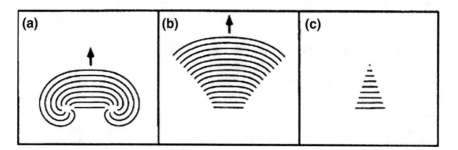

**Fig. 8.16** Evolution of a wave fragment in the anisotropic Wiener–Rosenblueth model. In these simulations, $G_0 = A - B(\sin\alpha)^2$ with $B = 1$. **a** Development of a pair of spiral waves, $A = 1.5$, **b** a steadily traveling and expanding fragment, $A = 0.5$, and **c** a shrinking fragment, $A = -0.5$. Superpositions of consequent positions of the wave are displayed. Reproduced from [23]

surface characterized by the slow "inhibitor" variable $w$. In the simulations, it was assumed that $D_x$ was constant, but $D_y$ depended on $w$ and took two values $D_1$ or $D_2$ with switching between them determined by the surface reconstruction fraction $w$ (hence, the anisotropy was getting weaker or stronger depending on the surface state). Under these assumptions, steadily traveling and expanding fragments could be seen. Moreover, their transformation into spiral waves or shrinking could also be observed.

While this analysis was motivated by the experiments with NO reduction on Rh, the same arguments are applicable for flat traveling wave fragments that are seen in the reaction of CO oxidation on Pt(110) whose existence region is indicated in the diagram in Fig. 8.11.

The present chapter concentrated on a single system, the oxidation of CO on a Pt(110) single crystal surface under isothermal ultrahigh vacuum conditions, for which the phenomena of spatio-temporal self-organization could be studied in most detail down to the atomic level. Similar investigations were performed with other crystal planes of platinum, as well as with other materials and other reactions [5].

Real catalysts consist usually not of extended crystal planes, but of small particles exposing different facets, and then novel effects through the propagation of reaction-diffusion waves across the boundaries between different planes can come into play. A good model for such systems is offered by the tip in a field ion microscope. In studies of the oxidation of hydrogen on platinum surfaces at low pressures and room temperature [24] it was observed that on a macroscopic Pt(100) surface the reaction reaches a steady state with the uniform distribution of the adsorbates. The field emitter tip, on the other hand, contained a region of the (100) plane with only 40 nm in diameter, and there the reaction exhibited sustained temporal oscillations. This effect was found to be associated with continuously varying distributions of the adsorbed species in the form of propagating waves, which were generated by coupling of reactions that occurred on adjacent crystal planes. It is felt that this effect is of a more general character in reactions catalyzed by small particles.

In a study with the CO oxidation reaction at elevated pressures and temperatures with small Pt particles inside a high-resolution electron microscope [25] it was found that the oscillatory reaction rate was synchronous with a periodic variation of the shapes of nanoparticles. This effect underlines the close coupling between reactivity and surface structure.

With the CO oxidation reaction on platinum another mechanism comes into play at higher pressures: Now the surface structure may switch between a metallic (reactive) and oxidized (less reactive) states. Theoretical modeling of the transition to oxide and its influence on the kinetics of CO oxidation revealed good agreement with the experimental observations [26].

Due to the pronounced temperature dependence of the reaction kinetics, even small deviations from isothermal conditions may cause significant effects on the dynamics. Such a situation can mainly occur with 'real' catalysts at elevated pressures. The direct coupling of kinetic oscillations to temperature variations and then in turn to mechanical deformations was demonstrated [27] with a thin (200 nm thick) platinum foil. Rate oscillations with a period of about 5 s at the $O_2$ pressure of $5 \times 10^{-3}$ mbar

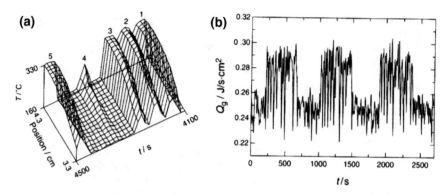

**Fig. 8.17** Thermokinetic effects during oxidation of propene on Pt. **a** Space-time diagram of the temperature profile. **b** Chaotic variation of the heat generated while the electric resistance is kept constant. Reproduced with permission from [28]

were accompanied by temperature variations of about 30 K. Such variations caused periodic mechanical deformations of the catalyst that were described as 'heartbeats'.

An example for very large temperature variations is shown in Fig. 8.17a for oxidation of propene on a 1 cm segment of a Pt ribbon which was heated electrically while its resistance (and thereby the average temperature) was kept constant. Irregular fronts are accompanied by chaotic kinetics. This is also reflected by the temporal variation of the total heat generated as shown in Fig. 8.17b [28].

# References

1. P. Hugo, Chem. Ing. Techn. **41**, 400 (1969)
2. P. Hugo, Berichte Bunsenges. Phys. Chem. **74**, 121 (1970)
3. H. Beusch, P. Fieguth, E. Wicke, Chem. Ing. Techn. **44**, 445 (1972)
4. M.M. Slinko, N. Jaeger, *Oscillating Heterogeneous Catalytic Systems* (Elsevier, Amsterdam, 1994)
5. R. Imbihl, G. Ertl, Chem. Rev. **95**, 697 (1995)
6. J. Rüstig, G. Ertl, unpublished (1981)
7. G. Ertl, P.R. Norton, J. Rüstig, Phys. Rev. Lett. **49**, 177 (1982)
8. T. Gritsch, D. Coulman, R.J. Behm, G. Ertl, Phys. Rev. Lett. **63**, 1086 (1989)
9. M. Eiswirth, G. Ertl, Surf. Sci. **177**, 90 (1986)
10. K. Krischer, M. Eiswirth, G. Ertl, J. Chem. Phys. **96**, 9161 (1992)
11. M. Takens, in *Dynamical Systems and Turbulence*, ed. by D.A. Rond, L.S. Young (Springer, Heidelberg, 1981), p. 83
12. M. Eiswirth, K. Krischer, G. Ertl, Surf. Sci. **202**, 565 (1988)
13. M. Eiswirth, G. Ertl, Phys. Rev. Lett. **60**, 1526 (1988)
14. K. Krischer, M. Eiswirth, G. Ertl, J. Chem. Phys. **97**, 307 (1992)
15. M. Ehsasi, O. Frank, J.H. Block, K. Christmann, Chem. Phys. Lett. **165**, 115 (1990)
16. H.H. Rotermund, W. Engel, M. Kordesch, G. Ertl, Nature **343**, 355 (1990)
17. M. Falcke, H. Engel, in *Spatio-Temporal Organization in Nonequilibrium Systems*, ed. by S.C. Müller, T. Plesser (Projektverlag, Dortmund, 1992)

18. S. Jakubith, H.H. Rotermund, W. Engel, A. von Oertzen, G. Ertl, Phys. Rev. Lett. **65**, 3013 (1990)
19. S. Nettesheim, A. von Oertzen, H.H. Rotermund, G. Ertl, J. Chem. Phys. **98**, 9977 (1993)
20. M. Bär, N. Gottschalk, M. Eiswirth, G. Ertl, J. Chem. Phys. **100**, 1202 (1994)
21. M. Bär, Räumliche Strukturbildung bei einer Oberflächenreaktion: Chemische Wellen und Turbulenz in der CO-Oxidation auf Platin-Einkristall-Oberflächen, PhD Thesis, Freie Universität Berlin, 1993
22. M. Bär, M. Eiswirth, Phys. Rev. E **48**, R1635 (1993)
23. F. Mertens, N. Gottschalk, M. Bär, M. Eiswirth, A.S. Mikhailov, R. Imbihl, Phys. Rev. E **51**, R5193 (1995)
24. V. Gorodetskii, J. Lauterbach, H.H. Rotermund, J.H. Block, G. Ertl, Nature **370**, 276 (1994)
25. S.B. Vendelbo, C.F. Elkjaer, H. Falsig, I. Puspitasari, P. Dona, L. Mele, P. Morana, B.J. Nelissen, R. van Rijn, J.F. Creemer, P.J. Kooyman, S. Helveg, Nat. Mater. **13**, 884 (2014)
26. P.A. Carlsson, V.P. Zhdanov, B. Kasemo, Appl. Surf. Sci. **239**, 424 (2005)
27. F. Cirak, J.E. Cisternas, A.M. Cuitino, G. Ertl, P. Holmes, I.G. Kevrekidis, M. Ortiz, H.H. Rotermund, H. Schunack, J. Wolf, Science **300**, 1932 (2003)
28. G. Philippou, F. Schulz, D. Luss, J. Phys. Chem. **95**, 3224 (1991)

# Chapter 9
# Electrochemical Reactions

Kinetic oscillations in an electrochemical reaction, observed by Wilhelm Ostwald already in 1899, served as the first example for oscillatory phenomena in chemical systems in Chap. 4. Subsequently similar effects were also intensively studied in the laboratories of Karl Bonhoeffer [1] and Ulrich Franck [2] in Germany.

As an example, Fig. 9.1 shows typical data, recorded during anodic dissolution of copper in a solution of hydrochloric acid, together with the scheme of the respective experimental set-up [3]. Temporal variations of the current and the potential were monitored by means of an oscilloscope. The following interpretation of the observed oscillations was given: This electrochemical reaction was accompanied by the development of surface coating with CuCl. At higher local current densities, $Cu^{2+}$ ions were formed in its pores and they caused the coating dissolution. The presence of an autocatalytic step was furthermore emphasized.

This interpretation followed a model that was first proposed by Bonhoeffer [4] and then elaborated [5] by Franck and FitzHugh (who was also one of the authors of the FitzHugh-Nagumo model of excitable media that we have discussed in Chap. 5). The model consisted of two coupled differential equations with the thickness of the coating and the potential for the formation of the coating (depending on pH) as the variables; it was solved using an analogue computer. Thus, a periodic limit cycle surrounding an instable stationary state could be identified. However, no clear connection to experimental observations was offered by this work.

The results of a more recent related study [6], where the periodic formation of a coating was absent, are shown in Fig. 9.2. Here, temporal variation of the electrode potential during the electro-oxidation of $H_2$ at a Pt electrode in the presence of a small concentration of $Cu^{2+}$ ions in perchloric acid was recorded under galvanostatic conditions (i.e. with the constant current density which served as a control parameter). Under an increase of the current density $j$, regular oscillations started with a small amplitude (signaling the occurrence of a Hopf bifurcation). Then the amplitude grew and its growth was followed by the period doubling. Finally, a transition to chaos characterized by the period-doubling scenario was observed. Subsequent

© Springer International Publishing AG 2017
A.S. Mikhailov and G. Ertl, *Chemical Complexity*, The Frontiers Collection,
DOI 10.1007/978-3-319-57377-9_9

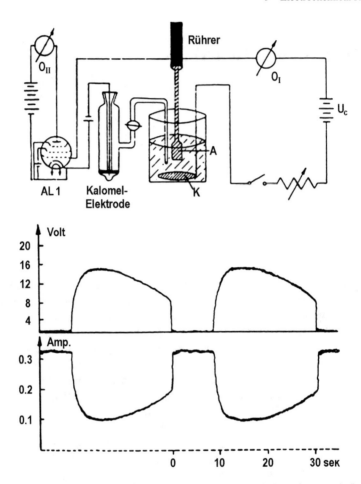

**Fig. 9.1** Observation of kinetic oscillations during anodic dissolution of copper in hydrochloric acid. *Above* the experimental set-up ("Rührer" = stirrer, "Kalomelelectrode" = saturated Calomel electrode). *Below* recorded oscillations of voltage and current. Reproduced with permission from [3]

investigations with the same system additionally revealed the emergence of spatiotemporal chaos with nonlocal spatial coupling effects [20].

An attempt for a more comprehensive theoretical description of temporal instabilities in electrochemical systems was undertaken by M. Koper [7]. His model is represented by an equivalent electrical circuit shown in Fig. 9.3. For the case of negative faradayic impedance, the theory yielded a universal phase diagram exhibiting both stationary and oscillatory behaviors of the system and the associated bifurcations. However, for the case of the galvanostatic control it predicted no oscillations, in contrast to the experimental data in Fig. 9.2.

Generally, temporal instabilities in electrochemical systems arise when the differential resistance in the current versus the double layer potential is negative.

**Fig. 9.2** The potential of a
platinum electrode as a
function of time during the
electro-oxidation of $H_2$ at
different current densities $j$.
Electrolyte: 1 M $HClO_4$,
$1.5 \times 10^{-4}$ M $Cu^{2+}$, $5 \times 10^{-5}$
M $Cl^-$. Reproduced from [6]

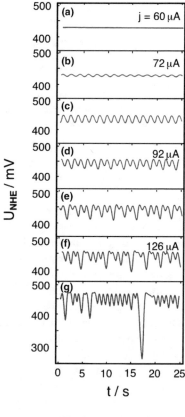

**Fig. 9.3** An equivalent
electric cell circuit
underlying the model of an
electrochemical cell [7].
$V$ is the fixed potential, $R_s$ is
the serial resistance, $E$ is the
electrode potential, and $Z_F$
is the Faradayic impedance

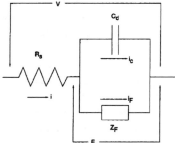

If spatial gradients of the concentrations of reacting species at an electrode surface
come into play, migration of ions under the influence of the potential of the double
layer becomes the dominant transport process, much faster than diffusion. Already
in 1891, Ostwald speculated [8] on the existence of such an effect which he called
*Chemische Fernewirkung* (chemical long distance action). Ostwald suspected some
analogy between the propagation of the excitation along a passivated iron wire,
which was studied by his coworker H. Heathcote [9], and functioning of a nerve.

This prompted the physiologist R. Lillie [10] to propose activation transmission in a passive metal as a model for the operation of nerves. In his own words:

> The propagation of the excitation wave is apparently dependent upon the bioelectric formed at the boundary between the active and inactive regions of the cell surface; that part of the local current which traverses the still inactive regions stimulates these electrically; the regions thus secondarily excited act similarly upon the resting regions next adjoining; the process repeats itself at each new active-inactive boundary as it is formed, and in this manner the state of excitation spreads continuously from active to resting regions. A wave of excitation thus travels over the surface of the element.

Subsequent detailed investigations by K. Bonhoeffer and W. Renneberg [11] of the activation propagation along a passivated iron wave confirmed this picture. They revealed the existence of a threshold and the decisive role of spontaneous changes of the potential at the surface.

A related process underlies the corrosion of stainless steel. There, a protective oxide layer prevents dissolution of the metal. This layer, however, usually contains defects where the film can locally break up. The onset of corrosion is characterized by the development of tiny seeds with a few micrometers size. Such a metastable pit is active for a few seconds and gets afterwards again passivated. This phenomenon is known as pitting corrosion. The transition to severe corrosion may occur suddenly and is accompanied by a sharp increase of the number of pits. It was suggested that this is a critical phenomenon characterized by autocatalytic formation of pits [12]: While a pit is activated, chemical reactions within it take place and aggressive products are released that diffuse into the surrounding solution. This causes damage of the protective layer in the neighbourhood of the pit and enhances there the probability for the formation of further pits.

The formation of individual pits was observed by optical microscopy, whereas the local thickness of the protective oxide layer was visualized by means of spatially resolved ellipsometry; both microscopic optical monitorings could also be performed in parallel at the same time [13]. Additionally, microcurrent spikes corresponding to an individual pit, while it was active, were recorded. The experiments confirmed [12] that the sudden onset of corrosion is accompanied by the formation and growth of regions with multiple metastable pits. Figure 9.4 shows the growth of the number of pits and the associated electrical current for a typical experiment; the exponential growth characteristic for critical phenomena is clearly seen. The spreading of such an area of pitting corrosion can be interpreted as the propagation of a transition front. The behavior was well reproduced by a stochastic model where the probability of pit formation depended on the local thickness of the protective layer, with active pits releasing aggressive products that diffused away and caused thinning of the layer around an active pit.

A general theory for spatial pattern formation in electrochemical systems was developed by G. Flätgen and K. Krischer in 1995 [14]. According to it, the potential distribution of the double layer affects migration currents on the whole interface between the electrode and the electrolyte. Therefore, such migration coupling is essential for the formation of spatial patterns whose evolution is determined by the local charge balance at the electrode. Hence, the reaction-migration rather than the

**Fig. 9.4** Microscopic observation of corrosion onset on stainless steel. *Above* the total number of pits on a logarithmic scale as function of time. *Below* the corresponding total electrical current. Reproduced from [12]

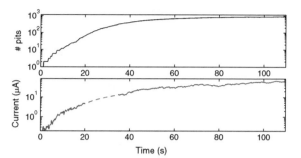

reaction-diffusion equations govern the dynamics in this case. An electrochemical reaction produces charges that become instantaneously balanced since no net charges can build up in the electrolyte. The condition of electroneutrality requires that for every surface element of the double layer the capacitive ($i_{cap}$) and the faradayic, or reactive, ($i_{reac}$) currents are balanced by the migration current ($i_{mig}$): $i_{cap} + i_{reac} = i_{mig}$. Taking into account these terms, one arrives at a general reaction-migration equation for the evolution of the double layer potential in space and time [14, 15].

There are important differences between the transport processes of diffusion and migration. The rate of diffusion is determined by the diffusion coefficient which is very similar for all species in solution and can mainly be varied by temperature. On the other hand, the conductivity of the electrolyte determines how rapidly inhomogeneities of the potential are adjusted by migration, and the conductivity can readily be varied by orders of magnitude. Moreover, there is a difference in the spatial range of coupling. The diffusive coupling takes place between the nearest neighbors and is therefore *local*, while a change in the state on an electrode induces practically instantaneous changes of the migration current in an extended surrounding area, so that the migration coupling is essentially *nonlocal* or *global*. The range of this extension depends on the distance between the working and the counter electrode and thus can also be varied in an experiment.

The operation of long-range coupling is well illustrated by triggering of potential waves in the potentiostatic electrochemical oxidation of formic acid on a platinum ring [16]. Under certain conditions, this system is bistable: it has a passive (OH-poisoned) and an active (high-current) states. A transition between them can be initiated by a local perturbation. The experimental set-up is schematically shown in Fig. 9.5. A platinum ring with the 42 mm outer diameter was used as the working electrode, a reference electrode was placed in the center of the ring and a Pt wire above them served as the counter electrode. To monitor the propagation of a potential front, 11 potential microprobes above the working electrode recorded the potential drop across the electric double layer. At position 12 a perturbation could be applied in the form of a positive pulse, causing local switching from the passive into the active state.

Experimental results are displayed in Fig. 9.6. In Fig. 9.6a, a pulse with the voltage of +3 V and the duration of 0.1 s was applied to the electrode 12 causing the system to

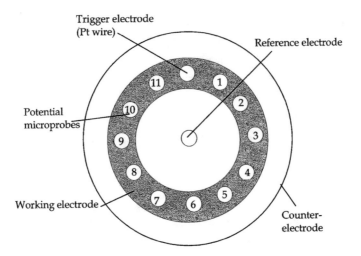

**Fig. 9.5** Experimental set-up for triggering of electrochemical potential waves

switch into the active state. As a result of such perturbation, two activation fronts were produced that propagated in opposite directions until they met after 0.2 s, leaving the system in the active state. (The passive state is associated with a strong potential drop across the double layer; hence a positive pulse will tend to reduce it.) This behavior can still be interpreted in terms of a local coupling. However, the situation becomes different in Fig. 9.6b. Here, a pulse with the negative voltage of −5 V and the 0.1 s duration was applied. The application of a negative pulse is expected to locally stabilize the double layer and initially the potential drop increased slightly at positions 1 and 11 next to the electrode 12. Then however a perturbation started to grow on the opposite side of the ring at the position 6 and it gave rise to two activation fronts. This clear example of remote triggering provides direct experimental evidence for the nonlocal nature of migration coupling. Such coupling appears instantaneous because of the quite different time scales of the reaction (about 0.1 s for the removal of the passivating OH layer) and of spreading of the electric field (about $10^{-9}$ s, determined by the speed of light).

Generally, nonlocal effects such as remote triggering are to be expected whenever a very fast coupling process operates in a system with slow local dynamics.

A strong difference in the rates of the two transport processes, migration and diffusion, opens a possibility for Turing patterns, as already mentioned in Chap. 3. A theoretical study of such effects, that was presented in 2000 by N. Mazouz and K. Krischer [17], revealed that in electrochemical systems with S-shaped current-potential characteristics under potentiostatic control the Turing patterns should indeed exist.

The experimental verification was performed soon afterwards [18]. The chemical system consisted of the reduction of periodate on an Au(111) surface in the presence of camphor; the lateral variation of the double layer potential on a thin film electrode could be imaged by means of a surface plasmon microscope [19].

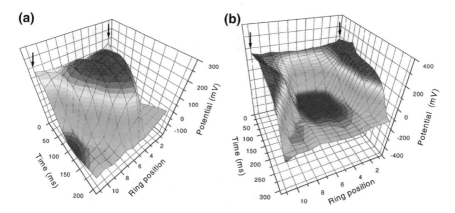

**Fig. 9.6** Evolution of the potential along the ring of electrodes with time for **a** local and **b** remote triggering. The perturbation is initially applied at the position 12. Reproduced from [16]

At negative potentials, the coverage by the adsorbed camphor is very low, while at the higher potentials a pronounced phase transition into a well-ordered dense adlayer occurs. In this potential range, the reduction of periodate becomes inhibited. Under potentiostatic conditions, such inhibition causes a shift of the double layer potential to the more cathodic (i.e. negative) values under which the camphor desorbs more readily. Hence, a negative feedback loop is created: Since the deposition of the camphor film is associated with a decrease of the current density for periodate reduction, the response of the system through the double layer potential will limit the growth of the camphor film. Thus, the camphor coverage will play the role of an activator, whereas the electrode potential will act as an inhibitor. The growth of the camphor film can be described by a reaction-diffusion equation. On the other hand, the condition of charge balance leads to a reaction-migration equation that describes spatial coupling between different locations on the electrode surface due to an inhomogeneous potential distribution. Since a change of the potential spreads much more rapidly than the adsorbate coverage, the requirements for the formation of Turing patterns are fulfilled.

Typical observed camphor coverage distributions at different potentials are shown in Fig. 9.7. They can be interpreted as single-spot Turing patterns whose size is controlled by the potential change. Patterns with two or three spots were however also observed depending on the reaction conditions. The patterns remained unchanged when the respective voltage was kept fixed. The size of the spots was determined by the difference of rate constants which could be varied to some extent by changing the conductivity of the electrolyte or the temperature.

The largest variety of electrochemical self-organization phenomena has been observed with the oxide layer on the n-type doped silicon electrodes during the dissolution under illumination [21]. Through an interplay of oxidation and etching of the oxide mediated by fluoride species, the system exhibits oscillations in

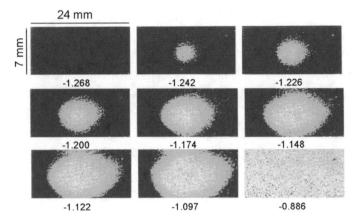

**Fig. 9.7** Single-spot Turing patterns. Camphor coverage distributions for different potentials under reduction of periodate at a gold electrode in a solution containing 5 mM camphor, 0.5 mM NaClO₄ and 0.5 mM NaIO₄. The numbers below each stationary distribution give the respective potentials. *Blue* regions are nearly camphor free, while *yellow* areas correspond to high coverage. Reproduced from [18]

the current and in the oxide layer thickness over a wide range of parameters. The electrochemical oxidation of silicon proceeds via the reaction

$$Si + 2H_2O + n_{VB}h^+ \rightarrow SiO_2 + 4H^+ + (4 - n_{VB})\,e^- \qquad (9.1)$$

where $n_{VB}$ is the number of electronic excitations induced by illumination across the valence band ($h^+$ denotes a hole and $e^-$ an electron). The created holes drive the charge transfer leading to nonlinear coupling between different positions on the electrode. On the other hand, the oxide becomes dissolved through the interaction with fluoride species. The state of the surface is monitored by optical ellipsometry.

Typical time dependences of the spatially averaged intensity of the ellipsometric signal are shown in Fig. 9.8. These comprise simple periodic oscillations (top), period-2 oscillations (middle) and irregular oscillations (bottom).

But even for regular integral oscillations (Fig. 9.9a) the lateral distribution is not uniform. This is seen in the two snapshots in Fig. 9.9b that are taken at the times indicated by two vertical lines in Fig. 9.9a. Remarkably, the patterns often consisted of several coexisting dynamic states. Thus, the surface was divided into distinct regions where different kinds of dynamics were observed. While in some regions synchronous oscillations took place, chemical turbulence was found outside of them.

Such a situation was encountered in 2002 by D. Battogtokh and Y. Kuramoto [22] in a theoretical study of the complex Ginzburg-Landau equation with nonlocal coupling. They have found that a system of identical oscillators splits into two kinds of domains. One is coherent and phase locked, and the other one is incoherent and desynchronized. Later on, this behavior was investigated in detail by D. Abrams and S. Strogatz [23] who have also coined a special term—the *chimera state*—to describe it:

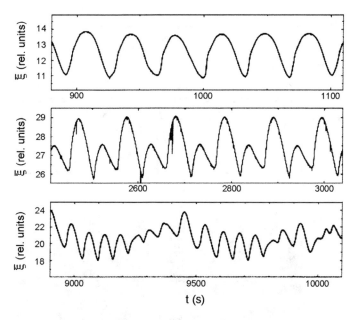

**Fig. 9.8** Time series for three typical measurements of the intensity of the ellipsometric signal (corresponding to the current density) during the oscillatory electrochemical oxidation of silicon under illumination: *top* periodic, *middle* period-2 and *bottom* irregular. Reproduced under permission from [21]

In Greek mythology, the chimera was a fire-breathing monster having a lion's head, a goat's body, and a serpent's tail. Today the word refers to anything composed of incongruous parts, or anything that seems fantastical...Nothing like this has ever been seen for identical oscillators. It cannot be ascribed to a supercritical instability of the spatially uniform oscillators, because it occurs even if the uniform state is stable. Furthermore it has nothing to do with the partially locked/partially incoherent states states seen in populations of nonidentical oscillators with distributed frequencies. Here, all the oscillators are the same.

While the chimera states were found afterwards in many theoretical and experimental studies, the mechanisms of their development are still being investigated. Recently, it was concluded that a clustering process to split the ensemble into two groups is necessary as the first step. Then the chimera states inherit their properties from the cluster states from which they originate [24, 25].

In this chapter, out attention was focused on electrochemical experiments with uniform systems. There were also many investigations by I. Kiss and J. Hudson with coworkers where specially designed electrode arrays were employed. As pointed out in Chap. 4, it was in these experiments that the Kuramoto synchronization transition was first observed in a chemical system [26] (see also [27]). By using electrode arrays, synchronization regimes for coupled chaotic oscillators were moreover explored yielding an excellent agreement with theoretical predictions [28]. We will return to the experiments with electrode arrays in the next chapter where design and control of self-organized systems are discussed.

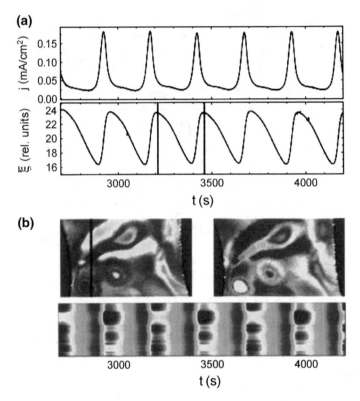

**Fig. 9.9** Coexistence of surface regions with different kinds of dynamics in the electrochemical oxidation of silicon under illumination. **a** Time series for the current and the ellipsometric intensity of the whole electrode. **b** Images of the ellipsometric state of the surface taken at the two times marked by *vertical bars* in (**a**). Below in this part the temporal evolution along the vertical line in the *left image* is displayed. *Brown* high, and *blue* low ellipsometric intensity. Reproduced under permission from [22]

# References

1.  K.F. Bonhoeffer, Z. Elektrochem. **47**, 147 (1941)
2.  U.F. Franck, Z. Naturforsch. **A4**, 378 (1949)
3.  K.F. Bonhoeffer, H. Gerischer, Z. Elektrochem. **52**, 149 (1948)
4.  K.F. Bonhoeffer, Z. Elektrochem. **52**, 24 (1948)
5.  U.F. Franck, R. Fitzhugh, Z. Elektrochem. **65**, 156 (1961)
6.  K. Krischer, M. Lübke, W. Wolf, M. Eiswirth, G. Ertl, Ber. Bunsenges. Phys. Chem. **95**, 820 (1991)
7.  M.T.M. Koper, Electrochim. Acta **37**, 1771 (1992)
8.  W. Ostwald, Z. Phys. Chem. **9**, 540 (1891)
9.  H.L. Heathcote, Z. Phys. Chem. **37**, 368 (1901)
10. R.S. Lillie, Science **68**, 51 (1918)
11. K.F. Bonhoeffer, W. Renneberg, Z. Phys. **118**, 389 (1941)
12. C. Punckt, M. Bölscher, H.H. Rotermund, A.S. Mikhailov, L. Organ, N. Budiansky, J.R. Scully, J.L. Hudson, Science **305**, 1133 (2004)

13. M. Dornhege, C. Punckt, J.L. Hudson, H.H. Rotermund, J. Electrochem. Soc. **154**, C24 (2007)
14. G. Flätgen, K. Krischer, J. Chem. Phys. **103**, 5428 (1995)
15. P. Grauel, J. Christoph, G. Flätgen, K. Krischer, J. Phys. Chem. B **102**, 10264 (1998)
16. J. Christoph, P. Strasser, M. Eiswirth, G. Ertl, Science **284**, 291 (1999)
17. N. Mazouz, K. Krischer, J. Phys. Chem. B **104**, 6081 (2000)
18. Y.-J. Li, J. Oslonovitch, N. Mazouz, F. Plenge, K. Krischer, G. Ertl, Science **291**, 2395 (2001)
19. G. Flätgen, K. Krischer, B. Pettinger, K. Doblhofer, H. Junkes, G. Ertl, Science **269**, 668 (1995)
20. H. Varela, C. Beta, A. Bonnefont, K. Krischer, Phys. Rev. Lett. **94**, 174104 (2005)
21. K. Schönleber, C. Zensen, A. Heinrich, K. Krischer, New J. Phys. **16**, 063024 (2014)
22. Y. Kuramoto, D. Battogtokh, Nonlinear Phenom. Complex Syst. **5**, 380 (2002)
23. D.M. Abrams, S.H. Strogatz, Phys. Rev. Lett. **93**, 174102 (2004)
24. L. Schmidt, K. Krischer, Phys. Rev. Lett. **114**, 034101 (2015)
25. L. Schmidt, K. Krischer, Chaos **25**, 064401 (2015)
26. I. Kiss, Y. Zhai, J.L. Hudson, Science **296**, 1676 (2002)
27. A.S. Mikhailov, D.H. Zanette, Y.M. Zhai, I.Z. Kiss, J.L. Hudson, Proc. Natl. Acad. Sci. USA **101**, 10890 (2004)
28. C. Zhou, I.Z. Kiss, J. Kurths, J.L. Hudson, Phys. Rev. Lett. **89**, 014101 (2002)

# Chapter 10
# Design and Control of Self-organizing Chemical Systems

When an architect designs a building, he develops individual blocks and then combines them into a desired structure. In a similar manner a chemical engineer works when a chemical factory is designed: The reactors for elementary processes are developed and then they are combined to form a production plant. To steer the operation of the plant into a different production mode, its manager would modify the processes in individual reactors and/or the network of production lines. Tacitly, such design and control are based on the assumption that individual structural parts and processes are independent and can be combined at will, perhaps under certain constrains.

This approach will not however work if a system is self-organized. Such systems are synergetic: their collective operation emerges from interactions between their elementary parts, but cannot be reduced to a combination of their effects. In this chapter, possibilities of design and control of self-organizing chemical systems will be discussed. Our analysis will be performed at a conceptual level and limited to simple examples.

As a starting point, we consider a system that collectively generates limit-cycle oscillations (but its individual elements are not necessarily oscillators themselves). We noted in Chap. 4 that the period, amplitude and shape of self-oscillations in a nonlinear dynamical system are uniquely determined by it and cannot be varied at will. This is in contrast with the situation for linear systems where oscillations with arbitrary amplitudes and different periods can be produced and, by superimposing them, various oscillation profiles can be reached.

A mathematician would say that the dynamics of a self-organizing system has an *attractor*. In the case of self-oscillations, the attractor is one-dimensional and represents a limit cycle. This means that, after a relaxation transient, the motion in the system is effectively characterized by only one variable, i.e., the oscillation phase, despite the fact that many interacting elements with a large number of degrees of freedom may be involved. The attractors can also have higher dimensionality, so that several collective variables, or, according to the German physicist Hermann Haken [1], *order parameters*, may be required to specify the motion on it. A system

© Springer International Publishing AG 2017
A.S. Mikhailov and G. Ertl, *Chemical Complexity*, The Frontiers Collection,
DOI 10.1007/978-3-319-57377-9_10

can be functional, i.e., having a predictable behavior that can be also controlled, only if the number of its order parameters is small enough. The dynamics of such systems proceeds on a low-dimensional manifold: one can imagine that they move on some hidden "rails". The motion along such rails is specified by a set of the order parameters.

Suppose that we want to control, by applying external perturbations, internal motions in a system with an attractive limit cycle. If our aim is to produce oscillations with a different amplitude or shape, it would be practically impossible to achieve it. Indeed, our perturbations would then need to be so strong that they destructively interfere with the interactions within the system responsible for the limit cycle. In other words, we would have to destroy the self-organization and to impose the dynamics that is very different from what the system likes to do itself. This would be as if we try to force a train to go off the rail-track.

The situation is however different if the period of oscillations is aimed to be controlled. While staying on its track, the train can be accelerated or slowed down and the time required to pass a given distance would thus be changed. Indeed, limit-cycle oscillators can be synchronized by an external periodic force. As a result, their oscillation period becomes equal to that of the external perturbation and the oscillation phase is locked to that of the periodic force. The initial oscillation phase is a free parameter because of the invariance with respect to time shifts.

It was noted by Turing that his stationary structures also have a free order parameter similar to the initial oscillation phase [2]. This parameter specifies the spatial position of the periodic structure (fixing the location of a maximum in it) and its existence is implied by the symmetry against space shifts. Since its position is arbitrary, a Turing structure can be shifted if an appropriate stationary perturbation is applied.

The same arguments hold for spatiotemporal self-organized patterns in homogeneous media. If, for example, we take a freely rotating spiral wave and move it as a whole to a different position, another valid solution of the reaction–diffusion equations will be obtained. In the experiments, weak perturbations would be enough to produce such an effect. In contrast to this, it is very difficult to change the shape of a spiral wave. Then one would have to force a self-organized structure from its hidden "rail track".

To control mechanical systems, physical forces can be used. For chemical systems, other kinds of perturbations have to be employed. In his early experiments, Zhabotinsky could achieve phase resetting in the BZ reaction by pulse injections of $Br^-$, $Ag^-$ and $Ce^{3+}$ ions [3]. To demonstrate entrainment of oscillations in the BZ reaction, its sensitivity to UV radiation was used by him [4].

In 1986, Lothar Kuhnert [5] has found that, if ruthenium is chosen as a catalyst, the BZ reaction becomes sensitive to visible light. Under illumination, $Br^-$ ions become produced in a photochemical cycle that act as an inhibitor for the autocatalysis. Thus, the excitability is decreased depending on the illumination intensity. The photosensitive modification of the BZ reaction has been broadly employed in the experiments on control of wave patterns.

In 1987, the resonance effect for spiral waves in the BZ reaction was observed [6]. In the experiment, the rotation period of a spiral wave was measured and then

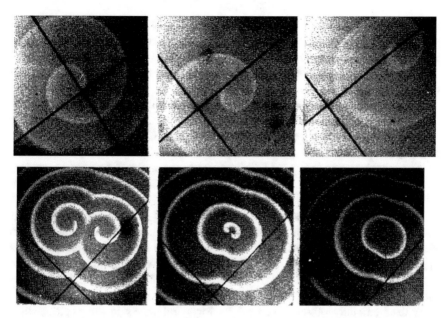

**Fig. 10.1** Resonance of spiral waves in the photosensitive BZ reaction. *Upper row* induced drift. Three subsequent snapshots are separated by intervals of 10 min, the total displacement of the spiral wave is 2 cm. *Lower row* induced annihilation of a pair of spiral waves. Reproduced from [6]

periodic uniform illumination with this period was applied. As a result of periodic forcing, the center of the spiral wave drifted with a constant velocity along a straight line (Fig. 10.1). The drift directions were opposite for clockwise and counterclockwise rotating spirals. When resonant periodic illumination was applied to a pair of counter-rotating spirals, they began to move one towards another and eventually their annihilation was observed.

Although it may appear unexpected, such resonance behavior can be easily understood. Suppose first that the medium is not uniform and a gradient of excitability along a certain direction exists. If there were no gradient, the tip of the spiral wave would have moved at a constant velocity along the circular core. When a gradient is applied, its motion is accelerated in the direction of the gradient during one half of the period and slowed down in the other half. Therefore, a drift of the spiral wave along the gradient should take place. Note that the drift is effectively caused in this case by the variation of the propagation velocity with the same period as that of the spiral wave. Thus, if, instead of a constant gradient, periodic uniform modulation of the excitability is applied and the modulation period coincides with that of the spiral wave, the same behavior should be observed. From such simple arguments it also follows that the direction of drift is opposite for a counter-rotating spiral wave. Moreover, the drift direction is determined by the illumination phase.

The mathematical theory of the resonance of spiral waves was constructed [6, 7] in the framework of the extended Wiener–Rosenblueth model. If periodic uniform modulation of excitability is applied at a frequency $\omega$ that is different from the natural rotation frequency $\omega_0$ of the spiral wave, the theory predicts that the center of the spiral wave will move at a constant velocity along a circle with the radius $R \propto (\omega - \omega_0)^{-1}$. This radius diverges as $\omega$ approaches $\omega_0$ and, under resonant conditions, drift at a constant velocity along a straight line takes place.

In Chap. 8, we have shown that spiral waves are also observed in the surface reaction of CO oxidation on Pt(110). Often such spiral waves are pinned by the surface defects and their centers are immobile (Fig. 8.12). However, some freely rotating spiral waves could also occur in this system, especially at the upper boundary of the excitable region in Fig. 8.11. To test whether a spiral is indeed freely rotating, periodic excitability modulation can be applied: while pinned spirals are not affected by such modulation, free spirals should start to move.

Such an experiment was indeed performed [8]. To modulate excitability, temperature was periodically varied and the spiral was monitored using photoemission electron microscope (PEEM). In this experiment, the modulation frequency was approximately twice as high as that of the spiral wave so that the conditions of the 1:2 resonance were satisfied. As shown in Fig. 10.2, the spiral drifted and was therefore free. The induced drift velocity was about 0.9 $\mu$m/s.

A detailed study of resonance effects for spiral waves in the photosensitive ruthenium-catalyzed BZ reaction was performed by Oliver Steinbock, Vladimir Zykov and Stefan Müller [9]. In these experiments, spiral waves were meandering in absence of forcing, i.e., their instantaneous rotation centers were additionally rotating along a circle. By applying periodic illumination near different $n{:}m$ resonances, motion of the tip along various hypocycloid trajectories was observed. These effects were qualitatively well reproduced in numerical simulations.

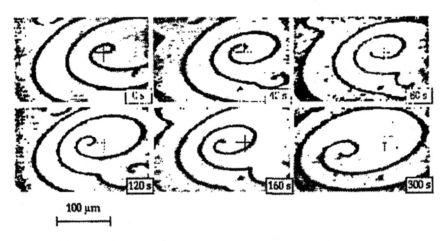

100 $\mu$m

**Fig. 10.2** Drift of a spiral wave induced by periodic modulation of temperature in CO oxidation on Pt(110). Reproduced from [8]

For numerical simulations of wave patterns in the photosensitive BZ reaction, a variant of the Oregonator model (7.9)–(7.10) is used [10]:

$$\frac{\partial u}{\partial t} = \frac{1}{\epsilon}\left[ u(1-u) - \frac{(bv+\phi)(u-a)}{u+a} \right] + D_u\nabla^2 u, \tag{10.1}$$

$$\frac{\partial v}{\partial t} = u - v + D_v\nabla^2 v, \tag{10.2}$$

where the variables $u$ and $v$ represent local concentrations of $HBrO_2$ and the active catalyst (ruthenium) ions. The diffusion constants of these two species are $D_u$ and $D_v$; if the catalyst is immobilized in a gel matrix, $D_v = 0$. The parameter $\phi$ is proportional to the light intensity. Under optical forcing conditions, it is varied periodically with time. As in the original model (7.9)–(7.10), dimensionless rescaled time is used in this description.

Effects of stationary illumination-induced excitability gradients on scroll rings in the BZ reaction were explored [11]. It was found that, depending on the direction and strength of the gradient, the contraction velocity of a scroll ring can be changed and expanding scroll rings can furthermore be produced. To understand such effects, one can recall that, in its cross-section, a scroll ring looks like a pair of counter-rotating spirals (cf. Fig. 6.12). When an excitability gradient is applied along the symmetry axis of the ring, the spirals start to drift along straight lines. If the radial (outward) component of the drift velocity is positive, the contraction rate of a scroll ring will be decreased and an expanding ring can also be obtained.

Because of the above-mentioned analogy between resonant forcing of spiral waves and their gradient-induced drift, similar phenomena should be expected if uniform resonant forcing is applied. The effect becomes particularly impressive for excitable media where, in absence of forcing, a scroll ring expands. As was shown in Fig. 6.13, such expansion leads to the development of the negative-tension turbulence of scroll waves. It has been demonstrated in numerical simulations and analytically confirmed that, by applying sufficiently strong periodic forcing, this kind of turbulence can be suppressed [12].

A typical simulation for an activator–inhibitor reaction–diffusion model is shown in Fig. 10.3. Here, the filament is visualized by the yellow curve. Additionally, activator distributions on three boundary planes are displayed. The simulation began at $t = 0$ with a small scroll ring. The ring started to expand and deform. Its filament stretched and became irregular, indicating that turbulence was formed (this process was previously demonstrated in Fig. 6.13). At time $t = 240$, uniform periodic forcing with the period close to that of a scroll ring was introduced. As a result, the filament started to shrink and to drift. At $t = 600$ in Fig. 10.3, only several short filament fragments remain. They continue to shrink, so that eventually the waves disappear and the medium goes into its rest state.

What will happen if an oscillatory reaction is periodically forced? Under continuous stirring, oscillation entrainments corresponding to different $n{:}m$ resonances should be observed, as described above. The situation is more complicated if stirring

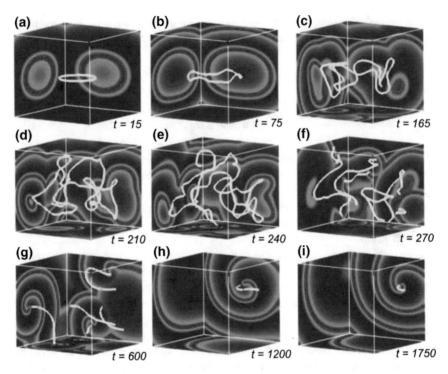

**Fig. 10.3** Numerical simulation of the action of periodic uniform forcing on the negative-tension turbulence of scroll waves. Periodic forcing is introduced at $t = 240$ and the turbulence becomes suppressed afterwards. The natural frequency of the scroll wave is $\omega_0 = 1.19$ and the forcing frequency is $\omega_f = 1.20$. Reproduced from [12]

is absent and diffusion of reactants caused by their concentration gradients takes place. To clarify it, experiments on optical periodic forcing of the oscillatory BZ reaction were performed [13]. Typical observed oscillatory patterns for different ratios $f_p/f_0$ of the modulation ($f_p$) and natural ($f_0$) frequencies are shown in Fig. 10.4.

The observations can be easily rationalized. Under the 1:1 resonance, synchronous oscillations of the entire system are seen whose phase is locked to that of the external force. The situation is different under the 2:1 resonance where the forcing period is half of the natural period. In this case, the oscillations can be locked either to even or odd numbers of forcing periods and therefore the system is effectively bistable. In the experiments, the system breaks into synchronous domains with the oscillations shifted by the forcing period. Such domains, separated by almost immobile fronts, are clearly seen in Fig. 10.4. Additionally, labyrinthine patterns were also observed in the vicinity of such resonance. When the forcing period was three times shorter, i.e. for the 3:1 resonance, the system exhibited triple stability and, respectively, three types of phase-locked domains were found (corresponding to black, gray and white shades in the last panel in Fig. 10.4). More complex cellular structures were found to emerge under the 3:2 resonance.

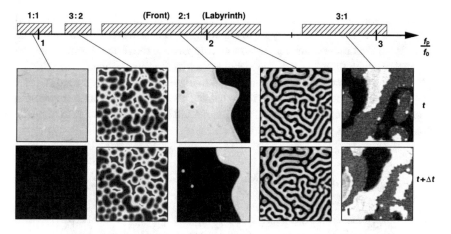

**Fig. 10.4** Patterns induced by optical periodic forcing of the oscillatory BZ reaction at different ratios $f_p/f_0$. Patterns are shown in pairs, one above the other, at times separated by $\Delta t = 1/f_p$, except for the 1:1 resonance where $\Delta t = 1/2f_p$. Reproduced with permission from [13]

Experiments with spatial periodic forcing of chemical Turing patterns were also performed [14]. In these experiments, the photosensitive chlorine dioxide–iodine–malonic acid (CDIMA) reaction was used. A continuously fed unstirred open reactor with the catalyst immobilized in a gel layer was employed. For spatial periodic modulation, the image of a mask with hexagonal patterns of a well-defined wavelength was projected at varying intensities above on the gel. The experimental results are reproduced in Fig. 10.5.

Without illumination (the first images in Fig. 10.5a, b), a Turing pattern of spots spontaneously formed. It did not however possess a perfect hexagonal symmetry and included many defects. Therefore, its Fourier spectrum consisted of a circle instead of individual peaks. The wave number distribution had a maximum at the wavelength $\lambda_0 = 0.53$ mm. In the first experiment (Fig. 10.5a), the spatial period of the projected mask was almost the same ($\lambda_f = 0.54$ mm). Under such resonant conditions, the main effect of illumination was that, as forcing was increased, the defects disappeared, the pattern became regular (as evidenced by the Fourier spectrum) and spot positions were locked to those in the mask (the second and the third images in Fig. 10.5a). A different behavior was observed away from an exact resonance, at $\lambda_f = 0.80$ mm. At intermediate intensities, restructuring began that finally resulted in a regular pattern with the spatial period imposed by the mask (the last image in Fig. 10.5b). Thus, the Turing pattern was effectively entrained.

In the above examples, order parameters were determined by the symmetry of a system. For a Turing pattern, the invariance with respect to rotations and translations implies that the order parameters should be the angular orientation of the pattern and its position on the plane. For a spiral wave, the order parameters specify the location of its center and the rotation phase. Such order parameters take arbitrary values and have no intrinsic dynamics.

Generally, however, order parameters are dynamical. The hidden rails on which a system moves are not flat, they form a landscape. It is however important that the dynamics of the order parameters is slow as compared to the rest of the variables that are "enslaved" by them [1], i.e., the system exhibits the separation of time scales.

Because the order parameters change only slowly, relatively weak perturbations are sufficient to control them. The force needed to move a train along its track up the slope is still much less than the force that would have been required to move the train over an arbitrary trajectory off the rails.

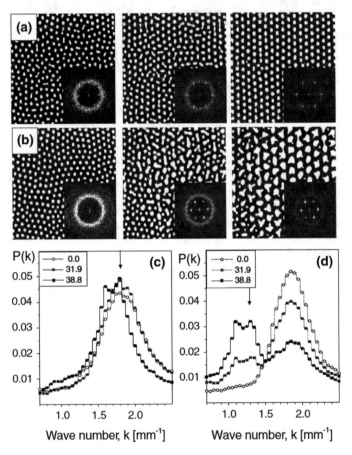

**Fig. 10.5** Spatial periodic forcing of Turing patterns in the CDIMA reaction **a** under resonance conditions ($\lambda_0 = 0.53$ mm, $\lambda_f = 0.54$ mm) and **b** off the resonance ($\lambda_0 = 0.53$ mm, $\lambda_f = 0.80$ mm). The *first column* shows patterns before the illumination, the *second* and *third columns* correspond to the light intensities of 31.9 and 38.8 mW/cm$^2$. Size of the pattern snapshots is $10 \times 10$ mm. The *insets* show the 2D Fourier spectra. **c, d** Wave number distributions averaged over the azimuthal angle for patterns shown in **a** and **b**. The *arrows* indicate the forcing wave number. Reproduced with permission [14]

For instance, an order parameter of the scroll ring is its radius. It is not however conserved: the scroll rings in the BZ reaction typically shrink. But they change their size on a longer time scale than, e.g., the rotation period of the scroll. Therefore, relatively weak spatial gradients are enough to prevent the collapse of a scroll ring [11]. Using weak periodic temporal forcing, spontaneous elongation of the filaments, and thus the turbulence, can be suppressed [12].

Obviously, the new dynamics cannot be just imposed "by brute force"—that would have meant that we destroy self-organization and act against it. Instead, what we need is to gently modify the system so that the self-organization processes within it are changed and the required structures thus naturally emerge.

Of course, this can be always done by changing the composition of the system, so that, for example, new physical interactions or chemical reactions are introduced. There is however also a more refined way of doing this where the use of artificial *feedbacks* is involved.

The basic scheme of a feedback loop is shown in Fig. 10.6. Information about the states of the elements of a system is collected and then used by a control device to generate a signal (or a set of signals). These signals are applied back to the system modulating its parameters in a prescribed way. The feedback is *global* if the information from all elements is included into a single control signal which is applied back in a uniform way.

The idea of using a negative feedback to stabilize a stationary state is very old: such a method was already used in the eighteenth century to control the operation of a steam engine. Norbert Wiener noted that various such built-in feedbacks are used by living biological systems to maintain their physiological steady states [15]. Later on, the Lithuanian physicist Kestutis Pyragas has suggested to apply time-delayed feedbacks for the stabilization of periodic orbits in systems with chaotic dynamics [16].

It should be stressed that, below in this chapter, feedbacks shall be employed in a different way. The aim is not to stabilize some already existing, but unstable structures. Instead, by means of an appropriate feedback, new kinds of self-organized structures in a system will be induced.

As an example, we consider a simple activator–inhibitor model where, by using a feedback, traveling spots could be obtained [17]. The evolution equations for local concentrations $u$ and $v$ are

**Fig. 10.6** The feedback loop

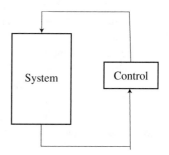

**Fig. 10.7** Stationary (*left*
$\sigma = 1$) and traveling (*right*
$\sigma = 0.0975$) spots. The
*color* is used to visualize the
inhibitor distribution, the
*black* contour shows the
activator interface. The
feedback parameters are
$\alpha = 5$ and $s_0 = 0.03$.
Reproduced from [17]

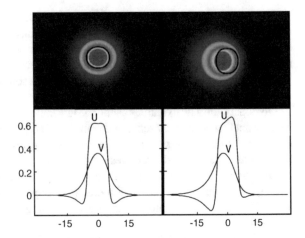

where $H(z)$ is the step function, $H(z) = 0$ for $z < 0$ and $H(z) = 1$ for $z \geq 1$.
The dimensionless time and length variables are chosen, with the characteristic time
and diffusion length of the activator set equal to one. The diffusion length $L$ of the
inhibitor was relatively large ($L = 8$). For the parameter values used in the study
($a = 0.275$ and $\mu = 3$), the system was bistable. Note that the parameter $a$ determines
the activation threshold.

The behavior of this system depends sensitively on the ratio $\sigma = L/\tau$. When this
ratio is sufficiently large, i.e., the inhibitor distribution can adjust relatively rapidly
to the changes in the activator distribution, stable spots are observed (Fig. 10.7,
left). They represent a special case of localized stationary structures discussed in
Chap. 3. The spot consists of a distinct core formed by the activator that is immersed
into a larger diffuse "cloud" where the inhibitor is localized. As the parameter $\sigma$ is
decreased and the inhibitor gets more inertial, the stationary spot becomes unstable.
Such instability is however subcritical. Numerical simulations reveal that, as its result,
the spot becomes transformed into a propagating excitation wave that extends to the
boundaries of the medium (as we noted in Chap. 6, excitation waves are possible in
bistable activator–inhibitor systems).

Suppose however that one wants to have a propagating localized structure, i.e.,
a traveling spot. With this purpose, an appropriate negative feedback is introduced
[17]. The activation threshold parameter $a$ is then controlled by the area $s$ of the spot,
$a = a_0 + \alpha(s - s_0)$. Thus, the threshold gets higher if the spot extends. In this model,
the spot area can be defined as

$$\frac{\partial u}{\partial t} = -u + H(u - a) - v + \nabla^2 u, \tag{10.3}$$

$$\tau \frac{\partial v}{\partial t} = \mu u - v + L^2 \nabla^2 v, \tag{10.4}$$

$$s = \int (u+v) d\mathbf{r}. \tag{10.5}$$

When such feedback is included, slowly traveling stable spots are obtained (Fig. 10.7, right). They have almost the same circular shape as a stationary spot, but the activator core is shifted with respect to the inhibitor cloud along the propagation direction (the feedback parameter $s_0$ is chosen approximately equal to the area of a stationary spot). Further away from the instability point, the spots become elongated and look like fragments of excitation waves. There is however no annihilation under collisions and scattering or fusion of the spots is instead observed.

The operation of the feedback can be rationalized in terms of the diagram in Fig. 10.6. A controlling device collects information about the activator and inhibitor concentrations $u$ and $v$ in all elements in the medium and sums its up to obtain the control signal $s(t)$. This signal is then used to modulate the threshold parameter $a$ that is common for all elements. Therefore, this is a global feedback scheme.

Note that such a scheme might be straightforwardly implemented in an experimental system where spots are observed. Then, an external controlling agent (such as a computer) should continuously monitor the area of a spot and generate the common control signal that changes the excitability of all elements. Essentially, the external feedback loop is used to modify the dynamics of the system, so that the desired structures (i.e., the traveling spots) are obtained.

The same result can be achieved without an external feedback if the system itself is appropriately modified. Suppose that, in addition to the activator $u$ and the inhibitor $v$, the system includes an additional species with the local concentration $w$. The extended system is described by equations

$$\frac{\partial u}{\partial t} = -u + H\left[u - a_0 - \alpha_w (w - w_0)\right] - v + \nabla^2 u, \tag{10.6}$$

$$\tau \frac{\partial v}{\partial t} = \mu u - v + L^2 \nabla^2 v, \tag{10.7}$$

$$\tau_w \frac{\partial w}{\partial t} = u + v - w + \mathcal{L}^2 \nabla^2 w. \tag{10.8}$$

If the diffusion length $\mathcal{L}$ of the new species is much larger than the system size, its distribution will be uniform. Moreover, if the kinetics of this species is very fast ($\tau_w \ll 1$), its concentration adjusts instantaneously to the current spatial distributions of $u$ and $v$ and we have

$$w = \frac{1}{S} \int (u+v) d\mathbf{r}, \tag{10.9}$$

where $S$ is the total area of the medium. It can be easily seen that, by choosing $s = Sw$, $s_0 = Sw_0$ and $\alpha = \alpha_w/S$, the action of the above global feedback can be indeed effectively reproduced.

**Fig. 10.8** Schematic
diagram of the
photochemical feedback
experiment. Reproduced
with permission from [18]

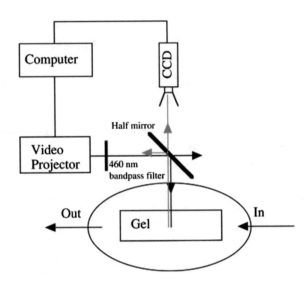

Localized traveling structures were investigated in the photosensitive BZ reaction
with a global feedback [18]. The scheme of the photochemical feedback experiment
is shown in Fig. 10.8. The reaction proceeded in an open flow reactor with the catalyst
present only inside a 0.1 mm thin gel layer. Images of waves were monitored with a
video camera. This data signal was analyzed by a computer that generated the control
signal which was applied back by means of a video projector. Comparing this set-up
with the general scheme shown in Fig. 10.6, we see that the role of a controller is
played by the computer in this case.

Because the inhibitor (i.e., the ruthenium catalyst) was immobilized in the gel,
stationary spots were not possible in such experimental system and the above discussed transition to slowly traveling spots could not be observed. Therefore, another
scenario that also yielded localized traveling structures had to be employed.

In Chap. 6, we have described early numerical simulations where the evolution
of an initial flat broken excitation wave with an open end was followed at different
excitabilities of the medium (Fig. 6.8). As we have seen, there exists a critical medium
excitability such that the flat wave fragment continues to propagate in its normal
direction retaining its length (Fig. 6.8b). If the excitability is decreased, the fragment
shrinks; it is is made stronger, the fragment growth and curls at its open end producing
a spiral wave. By optically varying the excitability of the ruthenium-catalyzed BZ
reaction, the illumination level that corresponds to such critical excitability could be
identified [18] and the experiments were performed in the vicinity of it.

The aim of the control was to obtain stable traveling wave fragments of a given
length. Because the thickness of a fragment is fixed and determined by the excitation
pulse width, the total excited area (i.e., the product of the length and the thickness)
had to be controlled. This area could be measured by analyzing the captured images
and counting the number $n_{exc}$ of pixels with the grey level above a certain threshold.

To implement the feedback, the illumination intensity $I$ of the projected light was computed according to the algorithm

$$I = an_{exc} + b,$$ (10.10)

with adjustable parameters $a$ and $b$. Because the excitability of the photosensitive BZ reaction is a decreasing function of the light intensity, a negative feedback loop was thus realized: if the filament grows, its excited area gets larger and, through the feedback, the excitability is decreased.

Figure 10.9 displays typical results. Depending on the feedback parameter $b$, stable traveling wave fragments of different lengths could be indeed obtained. They look similar to the traveling wave fragments observed in surface chemical reactions (Fig. 8.15). However, while the latter steadily expand, the length of such fragments is approximately conserved.

Qualitatively, the stabilization mechanism is clear: if the length begins to increase, the excitability is made lower and the wave shrinks; if it starts to shorten, the excitability is raised. A detailed theoretical description that allowed to reproduce this effect and predict the slightly curved shapes of the fragments was subsequently constructed [19]. In the experiments, adjustable illumination gradients could be additionally applied to guide the motion of short wave fragments along different preset trajectories on the plane [20].

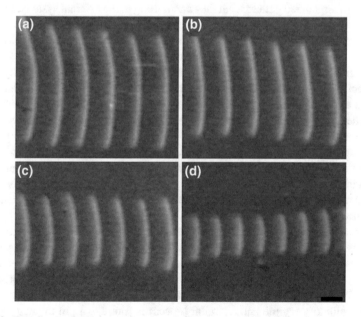

**Fig. 10.9** Traveling excitation wave fragments in the photosensitive BZ reaction with global feedback. By varying the feedback parameter $b$, wave fragments with different lengths are produced (**a–d**). In each *panel*, a superposition of images at 40 s intervals is displayed. The *scale bar* in *panel d* is 1.0 mm. Reproduced with permission from [18]

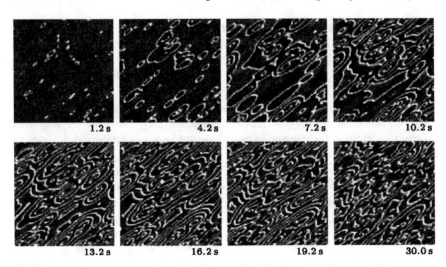

**Fig. 10.10** Spontaneous development of chemical turbulence from the uniform oxygen-covered initial state in the oscillatory CO oxidation reaction on Pt(110). A sequence of PEEM images with the size of $330\,\mu\text{m} \times 330\,\mu\text{m}$ is displayed. The reaction parameters are $T = 529\,\text{K}, p_{O_2} = 4 \times 10^{-4}$ mbar, and $P_{CO} = 12.3 \times 10^{-5}$ mbar. Reproduced from [22]

Thus, through a feedback, additional interactions between the elements forming a system are introduced. Such interactions are required for the emergence of the desired self-organized patterns, but they are missing in the original system in absence of the feedback. In that sense, inclusion of various feedbacks can also be viewed as leading to the modification of already existing systems and therefore also as an aspect of their design. When a computer-mediated feedback is present, as in the experiments we have just described, self-organizing systems with hybrid dynamics are effectively obtained. They combine intrinsic physical or chemical interactions with the *engineered interactions* through an external feedback loop.

The design aspect can be further illustrated by the experiments with the oscillatory CO oxidation reaction on Pt(110) where global delayed feedback was applied. In Chap. 8, we have noted that naturally present global coupling through the gas phase is essential for the development of uniform oscillations where periodic processes in even distant surface elements are synchronized. However, such naturally present global interactions may, under certain conditions, be not sufficient, and then chemical turbulence arises.

An example of the developing turbulence is shown in Fig. 10.10. Initially the surface is in the uniform oxygen-covered state. As the oscillations begin, small wave structures nucleate. Such structures grow and become irregular, so that eventually the catalytic surface is covered by many chaotically varying short fragments of spiral waves. To avoid confusion, it should be stressed that this behavior is characteristic for the oscillatory regime and a similar process is found, e.g., in the mathematical model of the complex Ginzburg–Landau equation (6.12). It is different from the development of turbulence in excitable media where turbulence first develops in the middle of an existing spiral wave and then spreads outwards (cf. Figs. 7.8 and 8.13).

In the experiments [21, 22], the feedback was used to suppress the turbulence and establish uniform oscillations. Additionally, new kinds of patterns could thus be obtained.

PEEM was employed to continuously image lateral distributions of adsorbed species on part of the surface. The feedback was introduced by making the instantaneous dosing rate of CO dependent on real-time properties of the imaged concentration patterns. The electronic signal controlling the dosing rate was generated by a computer that integrated the PEEM intensity over the whole image after a delay by a constant time $\tau_d$. The variation of the partial pressure of CO in the chamber followed the temporal modulation of the dosing rate with an additional delay determined by the residence time of gases in the pumped chamber. Thus, a global delayed feedback through the gas phase was introduced. The partial CO pressure $p_{CO}$ depended on the integral PEEM image intensity inside the observation window $I$ as

$$p_{CO}(t) = p_0 + \mu \left[ I(t - \tau) - I_0 \right], \tag{10.11}$$

where $p_0$ and $I_0$ are the partial CO pressure and the base level of the PEEM image intensity in absence of feedback (the PEEM intensity was defined in such a way that the darker areas had a higher intensity). The feedback was characterized by its intensity $\mu$ and the delay time $\tau$; both these parameters could be varied in the experiments.

The experimental results are summarized in Fig. 10.11. When sufficiently strong feedback was applied, suppression of turbulence and emergence of uniform oscillations could indeed be found (Fig. 10.11a). This transition is clearly seen in the spatiotemporal diagram along the line $ab$ in the middle panel in Fig. 10.11a. Additionally, the time dependences of the CO partial pressure $p_{CO}$ (black) and of the integral PEEM intensity $I$ (red) are shown below in this figure (the time scale is the same as in the spatiotemporal diagram). In the initial turbulent state, the oscillations are incoherent and therefore the integral PEEM intensity does almost not vary with time. During the synchronization transition, periodic oscillations in the control signal gradually emerged indicating the coherence onset. If the feedback intensity was somewhat lower, complete synchronization could not be achieved and the regime of intermittent turbulence was found instead (Fig. 10.11b). We return to the discussion of this regime below, when the respective theoretical work is discussed.

For other parameter values, phase clusters were observed (Fig. 10.11c). The surface became divided into large domains that oscillated out of phase. Examining the corresponding dependence of the partial CO pressure on time at the bottom of Fig. 10.11c, one can notice that the oscillations of this control signal have the period which is half the oscillation period within each of the domains. Thus, effectively the situation of the 2:1 resonance is realized, similar to that occurring under the optical forcing of the BZ reaction in Fig. 10.4. The difference is however that no external forcing with a shorter period is applied. Instead, the two groups of domains give contributions to the control signal that are shifted by half of the oscillation period.

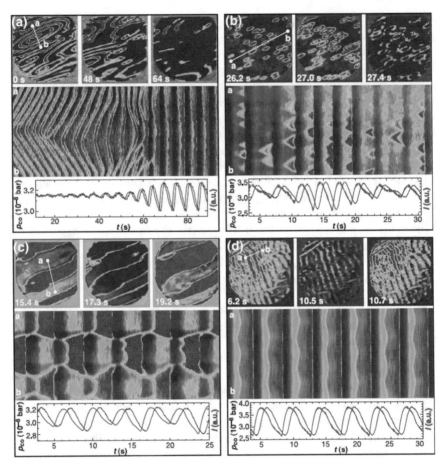

**Fig. 10.11** Spatiotemporal patterns observed in the experiments with global delayed feedback. **a** Suppression of spiral-wave turbulence, **b** intermittent turbulence, **c** phase clusters, and **d** standing waves. In each part, the *upper row* displays three subsequent PEEM images with a field-of-view of 500 μm in diameter. The images in the *middle row* are space–time diagrams showing the evolution along the line *ab* in the first image. The parameter values of temperature (K), oxygen partial pressure ($10^{-8}$ bar), base CO pressure $p_0$ ($10^{-8}$ bar), feedback intensity $\mu$ ($10^{-8}$ bar), and delay time $\tau$ (s) are, respectively: **a** 498, 10.0, 3.15, 0.8, 0.8, **b** 495, 10.0, 3.15, 2.0, 0.8, **c** 500, 10.0, 3.07, 0.6, 0.8, and **d** 505, 10.0, 3.30, 1.6, 0.8. Reproduced from [21]

Because oscillations are not harmonic, the sum of the two such contributions does not vanish and the signal varying with a twice shorter period naturally arises. Thus, in contrast to the experiments with the external forcing, the development of phase clusters in the surface chemical reaction is a self-organization effect.

Furthermore, the patterns of standing waves could be observed (Fig. 10.11d). In such patterns, the oscillations are overall synchronous, but the oscillation phase is periodically modulated in space. Similar patterns of standing waves were also observed in absence of the artificial feedback, as a result of the intrinsic global coupling through the gas phase [23]. Additionally, oscillatory cellular patterns could be found in the later experiments [22], the analogs of such cellular structures were also reported [24] in the experiments without a feedback.

Numerical simulations of feedback effects based on the model (8.9)–(8.11) were performed [25]. In these simulations, global delayed feedback was introduced assuming that partial CO pressure depends on the spatial mean $u_{av}$ of the CO coverage at a delayed time, i.e., $p_{CO}(t) = p_0 - \mu[u_{av}(t - \tau) - u_{ref}]$. Their results are summarized in the diagrams in Fig. 10.12. These data are for the one-dimensional system, but 2D simulations were also performed. The hysteresis was observed and different regimes could be reached depending on what initial condition was applied. Figure 10.12a shows the diagram based on the simulations starting from the state with the developed turbulence. The diagram corresponding to the simulations starting from a uniform initial state is displayed in Fig. 10.12b (note that this diagram covers only the smaller ranges of the control parameters $\mu$ and $\tau$). Synchronous uniform oscillations take place inside the light grey regions in the diagrams. Phase clusters (we return to them below) were observed inside the dark grey regions. Standing waves were found within the dashed area in the diagram in Fig. 10.12b (in 2D simulations, cellular structures could also be seen).

Figure 10.13 gives examples of possible wave regimes. The phase clusters (Fig. 10.13a) are clearly the same as in the experiments (Fig. 10.12b), with oscillations in the CO partial pressure having a period twice shorter than that of the oscillations within a phase-locked domain. In Fig. 10.13b, the spatiotemporal diagram corresponding to the intermittent turbulence is displayed. Such turbulent regime was found slightly below the synchronization boundary in the diagram in Fig. 10.12b. It was characterized by the appearance of localized bursts of turbulence on the background of uniform oscillations. Similar intermittent regimes can be found in the general model of the complex Ginzburg–Landau equation (6.12) extended to include a global feedback. For comparison, Fig. 10.13c shows bursts of turbulence in this

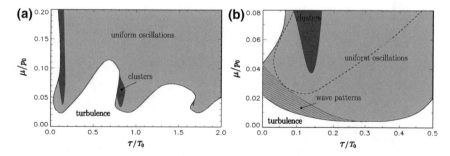

**Fig. 10.12** Synchronization diagrams based on numerical simulations started with **a** developed turbulence and **b** a slightly perturbed uniform state as initial conditions. Note the difference in the *axes scales* in **a** and **b**. The *dashed line* in **b** shows the synchronization boundary from **a**. $T_0$ is the period of uniform oscillations. Reproduced from [25]

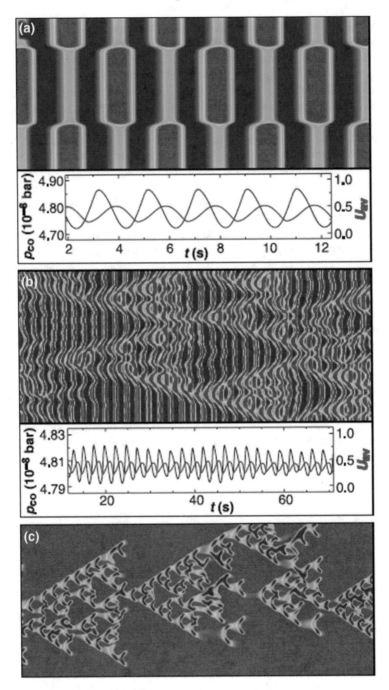

**Fig. 10.13** Spatiotemporal patterns in theoretical models with global feedback. **a** Phase clusters and **b** intermittent turbulence in the CO oxidation model. **c** Intermittent turbulence in CGLE with global instantaneous feedback. Reproduced from [21]

general model (the evolution of the local oscillation amplitude $|\eta|^2$ is visualized in this spatiotemporal diagram).

Control schemes based on local feedbacks were moreover designed and applied. In the experiments [26] for CO oxidation on Pt(110), concentration patterns were monitored by using ellipsomicroscopy for surface imaging. The camera images were continuously analyzed and processed in real time by a computer to generate a control signal. In turn, this signal was used to guide over the surface a laser beam that was focused to about 80 μm in diameter on the sample. The power of the beam was sufficient to locally raise the crystal temperature and thus to affect the reaction conditions. The light spot could be moved within 1 ms to any point on the surface with the precision of 5 μm by means of a pair of computer-controlled galvanometer mirrors.

A feedback loop became thus closed through a computer in this experimental system. As the authors remarked, "we demonstrate the implementation, through distributed sensing and feedback, of what one may term 'arbitrary evolution rules' - nonlocal interaction mechanisms on the surface that go beyond the nature of adsorption, desorption, surface diffusion, and reaction". By focusing an addressable laser beam to differentially heat the platinum surface, pulses and fronts, the building blocks of self-organized microscopic patterns, could be formed, accelerated, modified, guided, and destroyed at will [26] (see also [27, 28]).

Experiments with global feedback were also performed for the photosensitive oscillatory BZ reaction, using the illumination as the control signal, and various cluster patterns were found [29, 30]. Feedback control of spiral waves in the photosensitive excitable BZ reaction was explored [31]. A special feature of the latter experiments was that the control signal was not determined by integral properties of a pattern. Instead, the excitability of the medium was changed a little each time when the tip of the spiral wave was tangent to a certain line. Thus, a steady drift of the spiral along the selected line could be achieved.

Collective dynamics of interacting chemical oscillators can be purposefully modified, or *engineered*, by introducing appropriate feedbacks. To illustrate this, we consider the Kuramoto model (5.20) of interacting phase oscillators, i.e.,

$$\frac{d\phi_i}{dt} = \omega_i + \varepsilon \sum_{j=1}^{N} \Gamma\left(\phi_i - \phi_j\right). \tag{10.12}$$

Now we can notice that this model can be interpreted in a different way, with the interactions between the oscillators arising from global feedbacks.

Indeed, if the interaction function is $\Gamma(\phi) = \sin\phi$, this model can also be written as

$$\frac{d\phi_i}{dt} = \omega_i + \varepsilon \sum_{j=1}^{N} \sin\phi_i \cos\phi_j - \varepsilon \sum_{j=1}^{N} \cos\phi_i \sin\phi_j. \tag{10.13}$$

Introducing a global complex-valued control signal

$$Z(t) = \frac{1}{N} \sum_{j=1}^{N} e^{i\phi_j}, \tag{10.14}$$

the above equation can be reformulated as

$$\frac{d\phi_i}{dt} = \omega_i + \varepsilon \operatorname{Im}\left(Z e^{-i\phi_i}\right), \tag{10.15}$$

where $\operatorname{Im} z$ denotes the imaginary part of a complex number $z$.

Thus, equivalently, one can say that direct interactions between the oscillators are absent. Instead, there is an effective external controlling device that collects information about the phases $\phi_i$ of all oscillators and uses this information to generate the control signal $Z(t)$. After that, such signal is applied back to all oscillators according to the Eq. (10.15). Thus, a global feedback loop becomes established whose effect is exactly the same as that of direct interactions between individual pairs of oscillators in the Kuramoto model (5.20) with the interaction function $\Gamma(\phi) = \sin\phi$.

Suppose now that, instead of (10.14), a global signal

$$Z_k(t) = \frac{1}{N} \sum_{j=1}^{N} \left(e^{i\phi_j}\right)^k, \tag{10.16}$$

is generated and the Eq. (10.15) is replaced by

$$\frac{d\phi_i}{dt} = \omega_i + \varepsilon \operatorname{Im}\left(Z_k e^{-ik\phi_i}\right). \tag{10.17}$$

It can be easily checked that this global feedback would again result in the Kuramoto model, but with the interaction function $\Gamma(\phi) = \sin(k\phi)$.

By combining similar feedbacks of different orders $k = 1, 2, 3, \ldots$, *any* interaction function $\Gamma(\phi)$ can be reproduced. Because such interaction function determines the collective dynamics of the oscillator population, this dynamics can be effectively engineered by constructing the required feedbacks.

This idea was applied to engineer collective self-organized dynamics in the arrays of electrochemical oscillators [32]. An individual element in these experiments consisted of a nickel electrode in the $H_2SO_4$ solution at which the oscillatory dissolution reaction was taking place. The experimental set-up is shown in Fig. 10.14. The

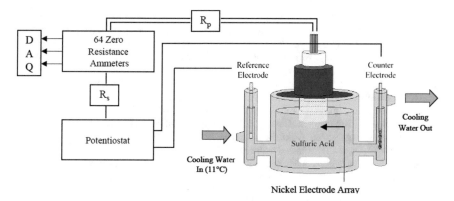

**Fig. 10.14**  The experimental set-up used for engineering of collective dynamics in electrochemical oscillator arrays. Reproduced with permission from [32]

electrode arrays with up to 64 elements were employed and the dissolution current was continuously monitored for each of them. This data was then used in real time by a computer to calculate the required feedback signal and send it back. The control signal was applied uniformly to all electrochemical oscillators through a potentiostat. The feedback signal was generated as a superposition of components of different orders $k$, each corresponding to a different power $i^k$ of oscillatory current variations in an electrode, similar to that described by Eq. (10.16).

By using this feedback set-up, systems with various kinds of self-organized collective dynamics could be effectively designed. Thus, a desired constant phase difference between two nonidentical chemical oscillators could be maintained and cluster regimes with irregular dynamics could be obtained where the clusters were sequentially rearranged. In a system of 64 coupled relaxation oscillators, desynchronization could be induced.

Remarkably, the experimental system in the study [32] was much more complex than the phase oscillators interacting according to the Kuramoto model, and even the equations describing a single oscillator were not available in this case. Nonetheless, the qualitative behavior was similar to the predictions made within the simple model (5.20).

# References

1. H. Haken, *Synergetics: An Introduction (Nonequilibrium Phase Transitions and Self-Organization in Physics, Chemistry and Biology)* (Springer, Berlin, 1977)
2. A.M. Turing, Philos. Trans. R. Soc. Lond. B **237**, 37 (1952)
3. V.A. Vavilin, A.M. Zhabotinsky, A.N. Zaikin, in *Biological and Biochemical Oscillators*, ed. by B. Chance (Academic, New York, 1973), p. 71
4. A.N. Zaikin, A.M. Zhabotinsky, in *Biological and Biochemical Oscillators*, ed. by B. Chance (Academic, New York, 1973), p. 81

5. L. Kuhnert, Nature **319**, 393 (1986)
6. K.I. Agladze, V.A. Davydov, A.S. Mikhailov, Pis'ma Zh. Eskp. Teor. Fiz. **45**, 601 (1987). English translation: JETP Lett. **45**, 767 (1987). Note that in the non-authorized English translation the term "helical waves" was used instead of "spiral waves"
7. A.S. Mikhailov, V.A. Davydov, V.S. Zykov, Physica D **70**, 1 (1994)
8. S. Nettesheim, A. von Oertzen, H.H. Rotermund, G. Ertl, J. Chem. Phys. **98**, 9977 (1993)
9. O. Steinbock, V.S. Zykov, S. C. Müller, Nature **366**, 322 (1993)
10. H. J. Krug, L. Pohlmann, L. Kuhnert, J. Phys. Chem. **94**, 4862 (1990)
11. T. Amemiya, P. Kettunen, S. Kádár, T. Yamaguchi, K. Showalter, Chaos **8**, 872 (1998)
12. S. Alonso, F. Sagues, A.S. Mikhailov, Science **299**, 1722 (2003)
13. V. Petrov, O. Qi, H.L. Swinney, Nature **388**, 655 (1997)
14. M. Dolnik, I. Berenstein, A.M. Zhabotinsky, I.R. Epstein, Phys. Rev. Lett. **87**, 238301 (2001)
15. N. Wiener, *Cybernetics, or Control and Communication in the Animal and the Machine* (MIT Press, Cambridge, 1948)
16. K. Pyragas, Phys. Lett. A **170**, 421 (1992)
17. K. Krischer, A.S. Mikhailov, Phys. Rev. Lett. **73**, 3165 (1994)
18. E. Mihaliuk, T. Sakurai, F. Chirila, K. Showalter, Faraday Discuss. **120**, 383 (2002)
19. V.S. Zykov, K. Showalter, Phys. Rev. Lett. **94**, 068302 (2005)
20. T. Sakurai, E. Mihaliuk, F. Chirila, K. Showalter, Science **296**, 2009 (2002)
21. M. Kim, M. Bertram, M. Pollmann, A. von Oertzen, A.S. Mikhailov, H.H. Rotermund, G. Ertl, Science **292**, 1357 (2001)
22. M. Bertram, C. Beta, M. Pollmann, A.S. Mikhailov, H.H. Rotermund, G. Ertl, Phys. Rev. E **67**, 036208 (2003)
23. S. Jakubith, H.H. Rotermund, W. Engel, A. von Oertzen, G. Ertl, Phys. Rev. Lett. **65**, 3013 (1990)
24. K.C. Rose, D. Battogtokh, A.S. Mikhailov, R. Imbihl, W. Engel, A.M. Bradshaw, Phys. Rev. Lett. **76**, 3582 (1996)
25. M. Bertram, A.S. Mikhailov, Phys. Rev. E **67**, 036207 (2003)
26. J. Wolff, A.G. Papathanasiou, I.G. Kevrekidis, H.H. Rotermund, G. Ertl, Science **294**, 134 (2001)
27. J. Wolff, A.G. Papathanasiou, H.H. Rotermund, G. Ertl, X. Li, I.G. Kevrekidis, Phys. Rev. Lett. **90**, 018302 (2003)
28. J. Wolff, A.G. Papathanasiou, H.H. Rotermund, G. Ertl, X. Li, I.G. Kevrekidis, Phys. Rev. Lett. **90**, 148301 (2003)
29. V.K. Vanag, L. Yang, M. Dolnik, A.M. Zhabotinsky, I.R. Epstein, Nature **406**, 389 (2000)
30. V.K. Vanag, A.M. Zhabotinsky, I.R. Epstein, J. Phys. Chem. A **104**, 11566 (2000)
31. J. Schlesner, V.S. Zykov, H. Brandtstädter, I. Gerdes, H. Engel, N. J. Phys. **10**, 015003 (2008)
32. I.Z. Kiss, C.G. Rusin, H. Kori, J.L. Hudson, Science **316**, 1886 (2007)

# Chapter 11
# Systems with Interacting Particles and Soft Matter

So far, phenomena of self-organization were discussed only for the systems where reacting species were not affected by mutual interactions apart from the reaction itself. This holds for dilute aqueous solutions such as in the BZ reaction, but also for surface and electrochemical reactions where the adsorbed species may be approximated by a 2D lattice gas. But, as we have pointed out at the beginning, motivation for studying phenomena of self-organization originated from attempts to understand how biological cells operate. These are, however, not filled with dilute aqueous solution in a chemical 'soup'. Instead, protein concentrations in the cytoplasm are so high that the biomolecules almost touch each other. Moreover, the cells include a cytoskeleton and are confined by biomembranes. Therefore, interactions between molecules should play an important role in biological cells.

In the case of surfaces, attractive interactions between adsorbed particles can give rise to the formation of ordered two-dimensional phases and also to spinodal decomposition, as illustrated, e.g., by Fig. 3.6. In the bulk, attractive interactions between constituents lead to condensed phases and eventually, if the interactions are strong enough, to solids in a crystalline state. In crystals, the mobility of atoms and, hence, also the chemical reactivity are strongly suppressed. There exists, however, a broad class of materials, collectively described as *soft matter*, for which condensation takes place, but the interactions are not strong enough to cause crystallisation. Among these are liquids, liquid crystals, polymers, gels, biological membranes, as well as Langmuir monolayers at a liquid/air interface.

Fundamental contributions to the theory of soft matter were made by the French physicist Pierre-Gilles De Gennes who was honoured in 1991 by the Nobel Prize "for discovering that methods developed for studying order phenomena in simple systems can be generalized to more complex forms of matter, in particular to liquid crystals and polymers."

The living systems consist essentially of soft matter and therefore the phenomena of self-organisation in them should be studied in order to answer Schrödinger's question '*What is life?*'.

© Springer International Publishing AG 2017

A.S. Mikhailov and G. Ertl, *Chemical Complexity*, The Frontiers Collection,
DOI 10.1007/978-3-319-57377-9_11

In this Chapter, we provide a short review of self-organization phenomena in the systems where both chemical reactions and energetic interactions between the particles are operating. After analyzing in more detail the model of reactive phase-separating binary solutions, we consider examples of inorganic soft matter formed by atomic adsorbates on solid surfaces. Then, we proceed to a discussion of non-equilibrium self-organization effects in Langmuir monolayers formed by surfactant molecules at a liquid/gas interface and in lipid bilayers representing biological membranes. Finally, the phenomena in bulk polymer systems and specifically in gels are discussed.

The interplay between chemical reactions and energetic interactions can be well illustrated by considering binary solutions with reactions between their components. First we summarize the behavior in absence of reactions as previously discussed in Chap. 3.

The equilibrium phase diagram of a mixture of particles $A$ and $B$ with repulsive interactions between $A$ and $B$ is shown in Fig. 3.3. In the middle of the phase separation region in the phase diagram, spinodal decomposition takes place. The theoretical description becomes simple near the critical point where the concentration difference $\phi$ between the two phases is small. The kinetic equation for $\phi$, formulated in 1958 by John Cahn and John Hilliard [1], is

$$\frac{\partial \phi}{\partial t} = M_0 \frac{\partial^2}{\partial x^2} \left[ -\alpha\phi + \beta\phi^3 - \kappa \frac{\partial^2 \phi}{\partial x^2} \right] \tag{11.1}$$

where the coeficient $\alpha$ is proportional to the deviation from the critical temperature $T_c$, i.e. $\alpha \propto (T_c - T)$. The initial stage of the decomposition process can be considered by linearizing this equation at $\phi = 0$. Then, the solutions are

$$\phi(x, t) = C \exp\left(ikx + \gamma_k t\right) \tag{11.2}$$

where (see Fig. 3.4)

$$\gamma_k = \alpha M_0 k^2 - \kappa M_0 k^4. \tag{11.3}$$

Initially, all modes with wavenumbers $k < \sqrt{\alpha/\kappa}$ grow and the fastest growth is at $k_0 = \sqrt{\alpha/2\kappa}$. Later, the characteristic length scale of the pattern increases and macroscopic domains occupied preferentially by particles $A$ or $B$ become formed.

What can change when the reaction $A \rightleftarrows B$ is introduced? If a system is closed, the reactions, like any other kinetic processes, will not affect its equilibrium state. Hence, the same macroscopic phase separation should be observed as the final state. The situation becomes however different if the system is open. Suppose, for example, that the forward reaction is accompanied by the consumption of a photon $\nu$, i.e. $\nu + A \rightarrow B$, and thus is energetically activated. Because energy is permanently supplied with the illumination and then dissipated into the thermal bath, non-equilibrium self-organization processes may take place. The theory of such processes was developed in 1995 by Sharon Glotzer, Edmund Di Marzio and Murugappan Muthukumar [2].

For simplicity, we consider only the case when the rate constants of both reactions are equal and they do not depend on the concentrations. Then, introduction of the reactions $\nu + A \rightarrow B$ and $B \rightarrow A$ into a mixture of particles $A$ and $B$ with repulsive interactions between them leads to an additional term in the Cahn–Hilliard equation (11.1), so that it becomes

$$\frac{\partial \phi}{\partial t} = M_0 \frac{\partial^2}{\partial x^2} \left[ -\alpha \phi + \beta \phi^3 - \kappa \frac{\partial^2 \phi}{\partial x^2} \right] - 2\Gamma \phi. \tag{11.4}$$

where $\Gamma$ is the reaction rate constant. Therefore, small periodic perturbations of the uniform state $\phi = 0$ with the wavenumber $k$ should now grow at the rates

$$\gamma_k = \alpha M_0 k^2 - \kappa M_0 k^4 - 2\Gamma. \tag{11.5}$$

Comparing Eqs. (11.3) and (11.5), one can notice that the instability of the uniform state should be suppressed (so that $\gamma_k < 0$ for all modes $k$) when the reactions are very fast, i.e. if the rate $\Gamma$ is high. When the reaction rate is gradually decreased, the instability occurs at $\Gamma_c = \alpha^2 M_0 / 2\kappa$ and, at the instability threshold, the mode with the critical wavenumber $k_0 = \sqrt{\alpha/2\kappa}$ begins to grow first. This critical wavenumber is the same as that of the fastest growing mode in the same system under spinodal decomposition in absence of the reactions. Note that the same instability is found if there are attractive interactions between particles $A$ (or particles $B$).

The subsequent nonlinear evolution is governed by Eq. (11.4). Numerical simulations reveal that the final state is stationary and represents a lamellar structure or an array of domains [2]. Near the instability threshold $\Gamma = \Gamma_c$, the final periodic stationary pattern has the same wavelength $\lambda_0 = 2\pi/k_0$ as the critical mode. Generally, the characteristic length scale $R_{eq}$ of the self-organized nonequilibrium pattern increases when the reaction rate constant $\Gamma$ is decreased, so that the macroscopic phase separation is recovered when $\Gamma \rightarrow 0$. For small reaction rates, $R_{eq} \propto (1/\Gamma)^3$, but the dependence becomes slower at the higher rates.

Experiments with binary polymer mixtures in the presence of photo-chemical reactions were subsequently performed [3, 4]. In absence of reactions, such mixtures were undergoing spinodal decomposition. Two types of chemical reactions, intermolecular photo-dimerization and intramolecular photo-isomerization, could be induced and controlled by irradiation with ultraviolet light. For both types, it was found that the decomposition becomes frozen and stationary spatial structures with intrinsic periodicities are established when the reactions are introduced. The behavior was however complex and elastic strains, developing in the blend, were also involved [4].

Not only bulk aqueous binary solutions or polymer blends, but also 2D monolayers, formed by adatoms on metal surfaces, represent systems with interacting particles. Because of attractive lateral interactions between the adatoms, a first-order phase transition leading to spontaneous separation of the surface into the domains with high and low coverages can take place. An example of spinodal decomposition of gold adatoms on Au(111) surface, monitored in real time with atomic resolution by scanning tunneling microscopy (STM), was previously shown in Fig. 3.6.

If we treat adsorbed atoms $A_{ad}$ and lattice vacancies $*$ as two different kinds of particles ($A$ and $B$), an adsorbate can be viewed as a binary mixture of them. Then, the condensation phase transition in the adsorbate can be viewed as spontaneous formation of domains where the adatoms or the vacancies prevail. In this case, the reaction $\nu + A \rightarrow B$ corresponds to photoinduced desorption, $\nu + A_{ad} \rightarrow *$, whereas the reverse reaction $* \rightarrow A_{ad}$ represents the process of adsorption from the gas phase. Note that, instead of photoinduced desorption, one can also consider a chemical reaction $S + A_{ad} \rightarrow * + P$ where a molecule $S$ from the gas phase hits the surface and reacts with an adatom $A_{ad}$, with the product $P$ immediately leaving the surface, so that a vacancy is formed.

For adsorbates, the characteristic wavelength of the patterns at the instability onset is estimated [5] as $\lambda_0 = \sqrt{r_0 L_r}$ where $r_0$ is the radius of attractive lateral interactions between the adatoms and $L_r = \sqrt{D/k_r}$ is their diffusion length with respect to a reaction or photo-desorption ($D$ is the surface diffusion coefficient and $k_r$ is the reaction/photodesorption rate constant). Since the interaction radius for electronically mediated lateral interactions is about the crystallographic lattice constant $l_0$ and the surface diffusion length $L_r$ can be about a micrometer, the characteristic wavelength $\lambda_0$ is expected to be of the order of $10\,nm$.

Thus, self-organized structures, resulting from an interplay between phase transitions and chemical reactions in adsorbates, are expected on nanoscales. In a fact, they represent clusters of adsorbed atoms whose typical size is controlled by the reaction rate. To describe such small structures, a mesoscopic theory for fluctuating adsorbate coverages was formulated [6] and applied to the adsorbates where, in addition to diffusion and reaction/photodesorption, thermal adsorption and desorption could take place [5]. The description was not limited to the vicinity of the critical point. Figure 11.1 shows an example of a structure obtained by numerical simulations. In this example, the linear size of the system is about 50 nm. While fluctuations are present, the basic lamellar pattern is clearly preserved.

Next, a system with two surface species $A_{ad}$ and $B_{ad}$ was considered [7]. The species were continuously supplied by adsorption from the gas phase and participated in the bimolecular surface reaction $A_{ad} + B_{ad} \rightarrow 2* + P$ with the reaction product immediately leaving the surface. Additionally, attractive interactions between particles $A$, and between $A$ and $B$, were introduced. It was found that *traveling* mesoscopic structures can develop in such a system [7]. The characteristic wavelength and the velocity of the structures were determined by the reaction rate.

Figure 11.2 shows the patterns obtained in a simulation. The fluctuations were strong and the waves were broken into short fragments forming an irregular pattern. Nonetheless, examining the time evolution in the central cross section (Fig. 11.2d), one can recognize that such fragments did not just fluctuate. These reactive atomic clusters traveled across the surface undergoing irregular variations of their shapes. The directions of motion of different fragments were random, and the fragments often collided. Merging and splitting of the fragments could take place. However, the magnitude of the propagation velocity of different fragments (determining the

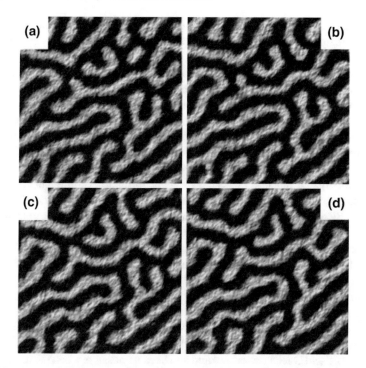

**Fig. 11.1** Mesoscopic self-organized structures. Numerical simulations for reactive phase separating adsorbates. The total size of the system is about 500 atomic lattice constants. Four consequent snapshots (**a**–**d**) are displayed. Reproduced from [5]

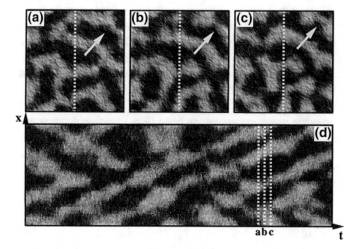

**Fig. 11.2** Mesoscopic traveling wave fragments. Spatial distributions of adsorbed particles $A$ at three consequent times are displayed (**a**, **b**, **c**). Temporal evolution along the central vertical cross section is shown in part (**d**). The length of the system is 555 atomic lattice constants. Reproduced from [7]

slope of stripes in Fig. 11.2d) did not significantly change. The total size of the simulated region was about 50 nm and the characteristic wavelength of a traveling self-organized structure was about 10 nm.

Special methods are needed for the observation of mesoscopic self-organized structures. Indeed, the photoemission electron microscopy (PEEM) has the resolution limit of a only few micrometers and cannot be applied in this case. In the experiments [8, 9], low energy electron microscopy (LEEM) with the lateral resolution of about 10 nm was employed and, moreover, the local composition of the structures was determined by using the X-ray photoemission electron microscopy (XPEEM) (these experiments were performed at Synchrotron Trieste).

The investigated reaction consisted of water formation on the Rh(110) surface in the presence of co-adsorbed Au (and also Pt) atoms. In this system, $H_2$ and $O_2$ molecules dissociate on the Rh surface, and H and O adspecies react to form water which immediately leaves the surface. Au does not directly participate in the reaction, but acts as an inhibitor, blocking the surface sites. Because the mixed Au+O state has a higher binding energy than the separated gold and oxygen phases, the formation of mixed Au and O islands was energetically favored. This process was however kinetically hindered by very low mobility of gold and oxygen adatoms. The water formation reaction dynamically removed the O adatoms from the surface, leaving empty sites that could be occupied by Au. A similar behavior took place if, in addition to Au, Pt atoms were present on the surface. Energetically, phase separation into mixed Au+Pd and O islands was favored and became possible when the reaction was introduced.

Two sequences of LEEM images in Fig. 11.3 show the reaction-induced evolution in the morphology of 0.55 monolayer of Au (left) or 0.75 monolayer of the mixture Au+Pd (right) deposited on Rh(110) surface. Before introducing the $H_2$ and $O_2$ reactants in the gas phase, the metal adatoms are uniformly distributed, and only step bunches on the Rh surface are seen ($t = 0$). When the reaction starts, lamellar metallic arrays (bright) develop and their wavelength begins to gradually increase. However, the spinodal decomposition process becomes later frozen and subsequently the morphology of the patterns changes only a little.

Such mesoscopic structures were also imaged at different reactant pressures while keeping other reaction parameters fixed [Fig. 11.4 (top)]. Their characteristic wavenumbers were determined by means of Fourier analysis and their dependence on the reactant pressure is plotted in Fig. 11.4 (bottom). The wavenumber increased with the pressure and the dependence approximately followed a power law with an exponent of 0.18.

These experimental results agreed qualitatively with general theoretical predictions [2, 5]. Lamellar structures with a definite characteristic wavelength are formed under reaction conditions. Their wavelength is controlled by the reaction rate (i.e., by the reactants pressure) and it grows according to a power law as the reaction rate is decreased. Further theoretical investigations were performed [10] for a model where two species arrive at the surface from the gas phase and react on it, forming a product that immediately leaves the surface. Additionally, a mobile promoter species was present on the surface which did not participate directly in the reaction, but could

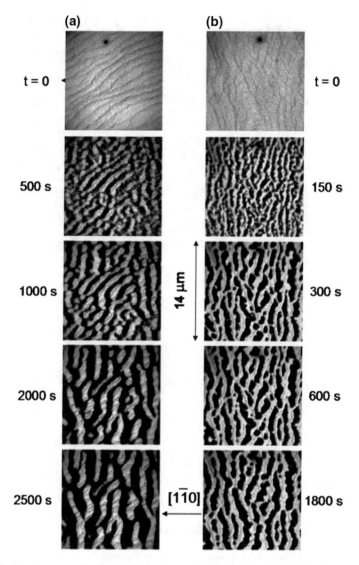

**Fig. 11.3** Development of self-organized mesoscopic Au (*left*) or Au+Pd (*right*) structures on the Rh(110) surface. LEEM images at indicated reaction times are displayed. **a** $\theta_{Au} = 0.55$ ML, reactant pressure $p = 5.3 \times 10^{-7}$ mbar, $p[H_2]/p[O_2] = 0.5$, $T = 780$ K; **b** $\theta_{Au+Pd} = 0.75$ ML, $p = 1.3 \times 10^{-6}$ mbar, $p[H_2]/p[O_2] = 0.8$, $T = 820$ K. Reproduced from [9]

inhibit (or enhance) it. Interactions between the promoter and one of the reactants were operating. The model was found to predict the formation of spatially modulated promoter distributions, similar to those experimentally observed.

In the above examples, nonequilibrium structures emerged as a consequence of an interplay between chemical reactions and a phase transition which was taking place in

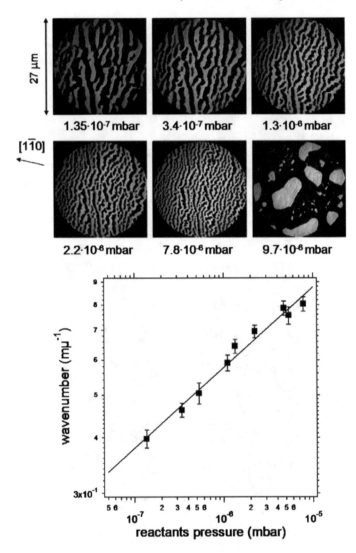

**Fig. 11.4** *Top* LEEM images of stationary self-organized mesostructures at different gas pressures of the reactants. (*bottom*) Log-log plot of the structures wavenumber along the [1$\bar{1}$0] direction. The *solid line* is a fit to a power law with an exponent of $0.18 \pm 0.02$. Reproduced from [9]

the system of reacting particles themselves. There are however also situations when reactions are coupled to a phase transition in a different physical system and such situations are typical for surface reactions as well. Adsorbed atoms often induce a reconstruction in the upper layer of the bulk according to a phase transition controlled by the adsorbate. In a turn, such reconstruction can change the reaction rate so that reciprocal coupling between a chemical reaction and a structural phase transition occurs.

This effect has played a principal role in the development of oscillations and excitability in the CO oxidation on Pt(110), as discussed in Chap. 8. The reconstruction does not proceed uniformly and, during it, the surface should consist of a mosaic of different phase domains. Such nanoscale domains are however too fine to be resolved with PEEM and, moreover, their dimensions are much smaller than the characteristic diffusion length for CO. Therefore, the surface could previously be characterized by a phenomenological variable representing the average local fraction of the reconstructed state. However, non-equilibrium self-organized nanostructures controlled by the chemical reaction can also arise in systems with the adsorbate-induced surface reconstruction and such effects were theoretically investigated too [11, 12].

Figure 11.5 illustrates the mechanism leading to the formation of a localized non-equilibrium structure, or a *self-organized nanoreactor*. Inside this structure, the surface is in the adsorbate-induced reconstructed state. The driving force is the gain in adsorption energy and, hence, the reconstructed surface region is attractive for the adsorbed molecules representing a potential well for them. Molecules diffusing across the surface are trapped by such regions and growing adsorbate islands are formed.

In absence of reactions, this process continues until the reconstructed regions become macroscopic or the whole surface becomes reconstructed. The situation is different if an energetically activated surface reaction (or the process of photo-induced desorption) is present. As a result of this reaction, adsorbed particles are removed from the surface. Because the reaction is activated, the differences in local adsorption energy between the reconstructed and nonreconstructed areas may play no role in determining the reaction rate constant.

An island grows due to the diffusion flux of molecules into it from the surrounding surface region. On the other hand, the reaction removes molecules from the island and slows down its growth. The competition between these two processes leads to the formation of stationary reactive islands [11].

The total diffusion flux of adsorbed particles into an island is proportional to its perimeter. However, the total rate of removal of particles from the island due to reaction is proportional to its area. For small islands, the incoming diffusion flux dominates and such islands grow. On the other hand, if an island is large, the reaction

**Fig. 11.5** A self-organized nanoreactor

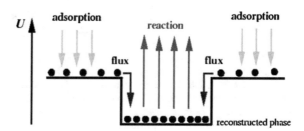

is prevailing and its size decreases. The two processes exactly balance each other at a certain radius that gives the size of the stationary localized structure. Such structures represent small self-organized chemical reactors. Indeed, the reaction essentially takes place only inside these reconstructed regions, where the adsorbate coverage is high.

Figure 11.6 (left) shows the theoretically determined dependence of the reactor radius $R_0$ on the reaction rate constant $k_r$. The lower dashed curve gives the radius of an unstable stationary structure, i.e. of the critical nucleus for the considered surface reconstruction phase transition. As the reaction rate is increased, the self-organized reactor becomes smaller and eventually disappears at the reaction rate constant comparable to the rate constant $k_{d,0}$ for thermal desorption. On the other hand, the reactor radius becomes large and diverges when $k_r \to 0$. Large self-organized reactors are however unstable. Figure 11.6 (right) shows the initial stage of the instability that eventually leads to extended structures covering the entire surface.

The smallest size of a reactor is of the order $\sqrt{l_0 L_r}$ where $l_0$ is the lattice constant and $L_r = \sqrt{D/k_r}$ is the diffusion length. Hence, such self-organized localized structures can persist down to nanoscales. They represent the analogs of spots in reaction-diffusion systems that we have discussed in Chap. 4. The minimal size of a spot in a reaction-diffusion system is however determined by the diffusion length. As we see, through coupling to an equilibrium phase transition, much smaller mesoscopic structures can be formed. Note that, for both kinds of structures, sufficiently strong local perturbations should be applied to create them.

Experimentally, self-organized traveling structures (shown in Fig. 11.2) and nanoreactors in surface reactions have not yet been observed. For traveling struc-

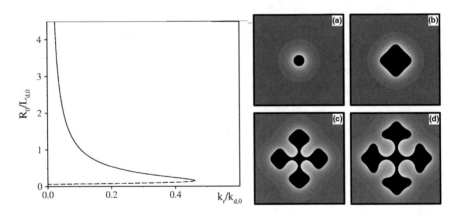

**Fig. 11.6** *Left* Dependence of the reactor radius $R_0$ on the reaction rate constant $k_r$. Here, $k_{d,0}$ is the thermal desorption rate constant and $L_{d,0} = \sqrt{D/k_{d,0}}$ is the diffusion length with respect to thermal desorption. The *dashed curve* corresponds to unstable solutions. *Right* The instability of a larger self-organized reactor. Four simulation snapshots **a–d** corresponding to the initial instability stage are shown. The local coverage is displayed in *gray scale*, increasing from *white* to *dark*. Reproduced from [11]

tures, the difficulty is that, to distinguish a traveling nanostructure from a merely fluctuating atomic cluster, its positions and shapes should be followed over time. In the case of self-organized nanoreactors, the problem is to distinguish them from equilibrium surface structures caused by heterogeneities and defects. Further discussion of self-organized nanostructures in adsorbate monolayers is given in the review [13].

Using the above examples, similarities and differences between equilibrium self-assembly phenomena and nonequilibrium self-organization in reaction-diffusion systems or in reactive soft matter can be discussed [14]. In Chap. 3, it was noted that, at thermal equilibrium, microphase separation can take place if, in addition to local interactions favoring phase separation, long-range interactions hindering the separation are present. Because of the latter interactions, macroscopic phase domains cannot be formed and a structure with alternating phase domains develops instead (see [15]). The development of such a structure is a self-assembly process in an equilibrium system. The structure corresponds to a minimum of the free energy and its properties are determined entirely by the parameters that may enter into this energy, such as the strengths and ranges of energetic interactions or the temperature. The kinetic coefficients may not however be involved. In contrast to this, self-organized Turing structures in reaction-diffusion systems, considered in Chap. 4, were of purely kinetic origin. They developed as a result of an interplay between reactions and diffusion and could be found in weak solutions where energetic interactions between the dissolved molecules do not play any role. The properties of such non-equilibrium patterns are determined by kinetic coefficients, such as reaction rates and diffusion constants.

The self-organized structures considered in this chapter occupy an intermediate position because they are determined by an interplay between energetic interactions and kinetic processes. Accordingly, their properties will generally depend both on the kinetic coefficients and the thermodynamic parameters of a system. Note that, despite the physical differences, mathematical descriptions of self-assembly and self-organization processes may sometimes be similar. For instance, the dependence (11.5) holds also in the case of equilibrium microphase separation (cf. Fig. 3.5). In a fact, it was shown by M. Motoyama and T. Ohta [16] that equilibrium microphase separation in diblock polymers is also described by Eq. (11.4).

Self-organization phenomena in reactive soft matter systems are expected to play a fundamental role in living cells. Indeed, only in such systems the self-organization can extend down to submicrometer and nanoscales. If only reactions and diffusion were involved, the minimal characteristic length scale of a pattern would have been limited by the diffusion length which is, for the typical biochemical reactions, of the order of a micrometer. But the cells are small: the bacteria have sizes of a few micrometers and the cells of macroorganisms are usually several tens of micrometers large. Hence, whereas reaction-diffusion patterns may still be significant on the scales comparable to the size of a cell, they cannot be responsible for the bulk of self-organization taking place inside it.

To illustrate principal mechanisms of self-organization in systems with interacting particles, catalytic reactions between small molecules adsorbed on metal surfaces have been used above. In biological cells, a different kind of systems—the organic soft matter—exists however. Therefore, self-organization phenomena in systems formed by biomolecules will now be analyzed.

It is convenient to start the discussion with Langmuir monolayers because they provide a bridge between the organic and inorganic systems. Such surface monolayers are formed by surfactant molecules disposed at a water–air interface. These molecules have hydrophilic heads, drawn inside the water, and hydrophobic tails which stay in the air. Depending on the surfactant concentration, equilibrium phases with different orientational ordering can be observed (Fig. 11.7).

Two-component Langmuir monolayers, representing binary surface solutions, can also be made. If interactions between different kinds of surfactant species are sufficiently strong, phase separation into surface domains predominantly occupied by one of the species take place. Furthermore, the two species can represent different conformations of the same molecule. This situation is characteristic for the derivatives of azobenzene where *trans* and *cis* isomers have largely different conformations. The *trans* isomer has an elongated shape with a tail, whereas the *cis* isomer is more compact, leading to different interactions between them.

Under illumination, transitions between these two isomers can be photo-induced. In the experiments by Yuka Tabe and Hiroshi Yokoyama [17], traveling orientation waves were observed in illuminated Langmuir monolayers in the tilted condensed state. The waves were accompanied by rotation of the azimuthal orientation direction. They propagated steadily at a velocity of about 50 μm/s and had a characteristic wavelength in the range of tens of micrometers. A typical example of the observed waves is shown in Fig. 11.8. The two images in this figure are separated by a second. A shift in the positions of the wave maxima (marked as A, B, C and D) is clearly seen.

A theoretical study for two-component phase-separating Langmuir monolayers with photo-induced transitions between the two components has been performed [18, 19]. In this model, traveling concentration and orientation structures with a definite wavelength and propagation velocity were found to exist. An example of such a pattern, similar to the experimentally observed orientational waves, is shown in Fig. 11.9. To improve the agreement with the experiment, polarization anisotropy had however to be additionally considered [20]. The detailed account of such experiments and of their theoretical modeling is given in the review [21].

Not only illumination can be used to bring Langmuir monolayers away from thermal equilibrium and to induce self-organization. In the experiments [22], the monolayers included a fraction (about 10%) of chiral molecules. These monolayers

**Fig. 11.7** Langmuir monolayers are formed by surfactant molecules disposed at the water–air interface. Different kinds of orientational ordering are possible for them

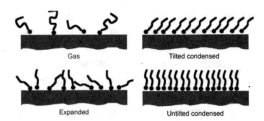

Gas                     Tilted condensed

Expanded                Untilted condensed

were at an interface between the air and glycerol. Inside the glycerol, water was dissolved and water vapor was also present in the atmosphere. The concentration of water molecules in the liquid was higher than in the air and therefore water was evaporating from the liquid through the surface monolayer into the air.

The compounds used in the experiments had a rod-like molecular structure with a chiral group attached, thus they looked like molecular propellers (Fig. 11.10). Under experimental conditions, these molecules were in the orientationally ordered, condensed phase. When a gradient of water concentration across the surface monolayer was existing, steady rotation of the azimuthal orientation was observed by means of the reflected-type polarizing microscope. Such coherent precession was very slow, with characteristic periods of several minutes [22]. Its frequency was directly proportional to the concentration gradient of water across the layer. The precession was

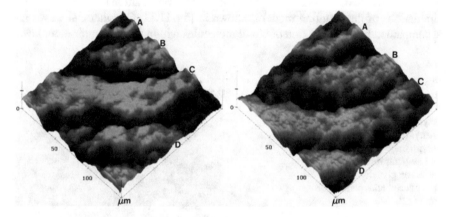

**Fig. 11.8** Orientational traveling waves induced by *trans-cis* photoisomerization in a Langmuir monolayer. The patterns are imaged using optical polarization (Brewster) microscopy. Two snapshots (*left* and *right*), separated by 1 s, are shown. Reproduced with permission from [17]

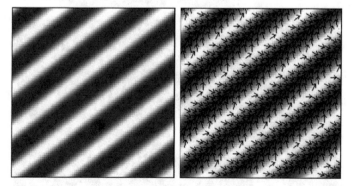

**Fig. 11.9** Numerical simulations of traveling orientational waves. The concentration (*left*) and orientation (*right*) fields are displayed. The local tilt is shown by the *grey* level and the azimuthal orientation is indicated by *arrows*. The stripes move in the diagonal direction. Reproduced from [18]

not uniform and various spatiotemporal patterns were observed. In a pacemaker-like pattern, the central region acted as a periodic source of waves that propagated towards the outer boundaries of the system [22]. The wavelength decreased as the waves propagated away from the center.

Y. Tsori and P.-G. De Gennes [23] have explained this phenomenon by pinning of the orientational field on the boundaries of a Langmuir monolayer. Indeed, if the orientation field rotates in the center, but is pinned at the boundaries, one should observe concentric waves which are regularly emitted at the central region and propagate outwards. Complex wave patterns were however also experimentally observed. Structures formed by several pacemakers with colliding wavefronts, making the system look "like a pond surface hit by falling raindrops", were reported [22] and rotating spiral waves could be furthermore seen [24]. To reproduce such complex patterns in the theory, coupling between local concentration of chiral molecules and the orientational field had to be included [25]. An example of a wave pattern found in numerical simulations of the resulting model is shown in Fig. 11.11. Orientational waves are accompanied by redistribution of chiral molecules within the Langmuir monolayer.

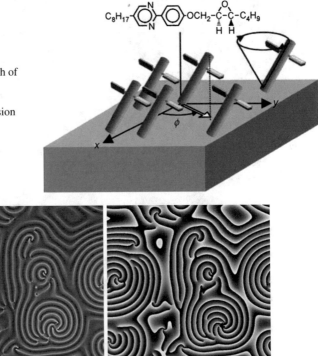

**Fig. 11.10** Chemical structure of (R)-OPOB {(2R)-2-[4-(5-octylpyrimidine-2-yl)-phenyl-oxymethyl]3-butyloxirane} and a sketch of its Langmuir monolayer spread on glycerol. Reproduced with permission from [22]

**Fig. 11.11** Numerical simulations of complex wave patterns in Langmuir monolayers with trans-membrane flows. Local concentration of chiral molecules (*left*) and the azimuthal orientation field (*right*) are displayed. Reproduced from [25]

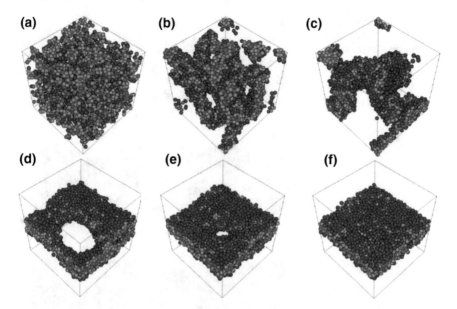

**Fig. 11.12**  Self-assembly of a biological membrane. A coarse-grained numerical simulation with the explicit solvent (its particles are not displayed). The initial state **a** represents a uniform mixture of lipids. Reproduced from [26]

Closely related to Langmuir monolayers are biological membranes that play a fundamental role in the living cells. Similar to Langmuir monolayers, they are formed by molecules—lipids—which have hydrophilic heads and hydrophobic tails. A uniform mixture of lipids in water is thermodynamically unstable and, through a self-assembly processes, a biomembrane representing a lipid bilayer becomes built (Fig. 11.12). Water is expelled from the center by the hydrophobic molecular tails. Like in Langmuir monolayers, different structural phases are possible in biomembranes. At relatively low temperatures, orientational ordering of lipids takes place (Fig. 11.13a). As temperature is increased, such order disappears and gives rise to a liquid phase (Fig. 11.13b). Under further temperature increase, water begins to penetrate inside the membrane, fluctuations grow (Fig. 11.13c) and the bilayer becomes disassembled. Inside the living cells, biomembranes are usually found in the liquid phase.

Biological membranes typically contain protein inclusions making up to 40% of the total membrane mass. Many inclusions, such as, e.g., ion pumps, actively change their conformations and effectively operate as molecular machines (to be discussed in the next chapter). Because of the coupling between the inclusions and the membrane, conformational changes in them induce local shape changes in the surrounding area of the membrane. As shown by Jacques Prost and Robijn Bruinsma [27], protein activity can result in the development of non-equilibrium fluctuations in the membrane shape. These predictions were confirmed [28] in the experiments

**Fig. 11.13  a–c** Structures of
lipid bilayers at different
temperatures (see the text).
Hydrophilic heads are *dark
blue*, terminal beads of
hydrophobic tails are *light
blue*. Reproduced from [26]

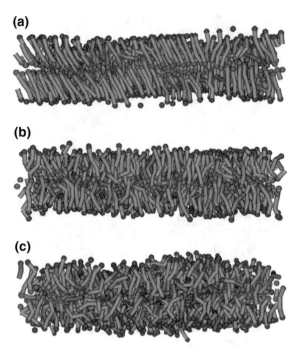

where light-driven proton pumps (the bacteriorhodopsin) were incorporated into the
phospholipid bilayer of giant vesicles.

Microscopic imaging of spatial distributions of proteins incorporated in biomembranes is difficult because the membranes are flexible and their shapes fluctuate. To

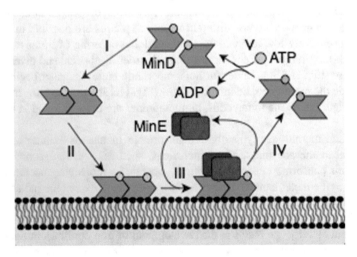

**Fig. 11.14**  Scheme of the reaction involving membrane-bound proteins (see the text). Reproduced
with permission from [29]

**Fig. 11.15** Spiral wave observed by laser scanning confocal microscopy for a reaction on a supported membrane. Reproduced with permission from [29]

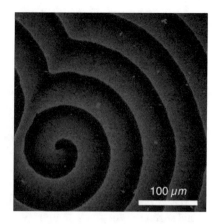

avoid such complications, it is convenient to work with supported membranes. Such membranes are produced by depositing lipid bilayers onto solid surfaces. In this way, flat bilayers are obtained and their accurate optical imaging can be performed. The supported membranes retain their functionality. Particularly, proteins can arrive from the water solution above the membrane and become adsorbed on it. The adsorbed proteins can laterally diffuse and react, with the products desorbing into the liquid phase. The entire setup resembles therefore the experiments with surface reactions on metal surfaces. In the latter experiments, the adsorbed particles are however atoms or small molecules that come from the gas phase.

Using supported lipid bilayers, self-organization processes under chemical reactions in biomembranes were experimentally explored [29]. In these experiments, a membrane reaction involved in the process of cell division was in vitro reproduced (Fig. 11.14). Proteins MinD, dissolved in water above the membrane, dimerize in the presence of adenosine 5'-triphosphate (ATP) (stage I in Fig. 11.14) and bind to the supported lipid membrane (stage II). Another type of proteins (MinE), also present in the solution, binds to the membrane-bound MinD dimers (stage III) and induces the hydrolysis of ATP. Subsequently, both proteins detach from the membrane and dissociation of MinD dimers occurs (stage IV). The hydrolysis product ADP becomes replaced by another ATP molecule (stage V), and the cycle is repeated again.

Proteins were labelled to make them fluorescent and, to observe membrane distributions of labelled molecules, laser scanning confocal microscopy was employed. Initially, the solution containing only MinD proteins was prepared and the reaction was started by adding MinE molecules to it. After the addition of MinE, planar surface waves were formed on the membrane. They moved in a definite direction across the membrane and persisted for several hours. The waves were composed by periodic bands of MinD and MinE, separated by troughs devoid of proteins. Depending on the concentration of MinE, their velocity was between 0.28 and 0.8 μm/s and the wavelength varied from 100 to 0.55 μm. Steadily rotating spiral waves were also observed (Fig. 11.15). The wave patterns could be described [29] by a reaction-diffusion model for concentrations of proteins in the solution and on the membrane.

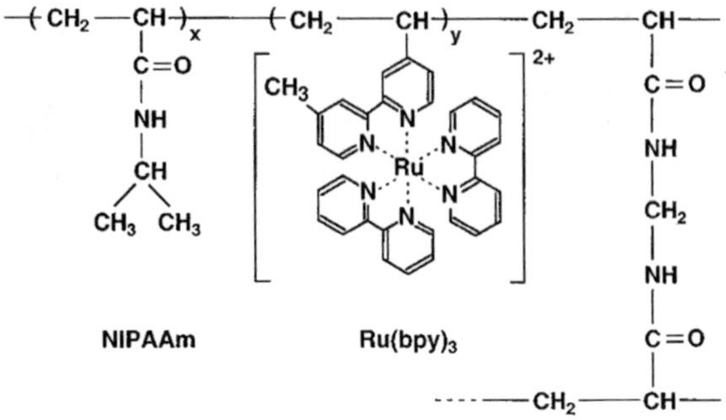

**Fig. 11.16** The chemical
structure of the ionic gel
used in the experiments [30]

NIPAAm                          Ru(bpy)₃

The similarity between these experiments and catalytic chemical reactions on metal surfaces is obvious. In both systems the reactants arrive at the surface from the bulk and are released into it, with the reaction taking place only on the surface itself. However, now the surface is formed not by a metal, but by the supported membrane, and the bulk is a water solution, not a gas.

Note also that the membrane played a passive role in such an experimental set-up, i.e. it provided a medium where the self-organization based on reactions and diffusion took place. This is different from CO oxidation on Pt(110) where a structural phase transition on the supporting Pt surface is induced by the adsorbate and, in a turn, affects the reaction course. It is however also possible that, in soft matter systems, reactions lead to phase transitions in the supporting medium and such situations are encounted in the experiments with gels.

The materials known as gels represent elastic networks of cross-linked polymer fibers. Such networks are rather open and much water can be contained within them. The water may include various chemicals and reactions between them can freely take place inside a gel. Hence, a gel provides a medium where self-organization processes based on reactions and diffusion can take place. In a fact, many experiments with the Belousov–Zhabotinsky reaction described in Chap. 7 were perfomed by using gels.

It is important that swelling-deswelling phase transitions are possible in a gel. When temperature is decreased, such material goes into a more compact deswelled structure where less water is contained. The critical temperature can depend on the concentrations of chemical substances dissolved in water and, furthermore, reactions between them can be affected by the physical structure of a gel. Thus, reciprocal coupling between chemical reactions and physical phase transitions can take place.

R. Yoshida, E. Kokufata and T. Yamaguchi synthesized [30] an ionic gel consisting of cross-linked N-isopropy lacrylamide (NIPAA) chains to which ruthenium *tris*(2,2′-bipyridine) [Ru(bpy)₃] was covalently bound (Fig. 11.16) and the Ru(bpy)₃ moiety acted as a catalyst for the Belousov–Zhabotinsky reaction. In their experiments, the gel was immersed into an aqueous solution containing all reactants of the BZ reaction, except for the catalyst itself. Hence, the catalytic reaction step could proceed only within the polymer fibers of the gel. When the reaction takes place, this atomic group alternates between two differently charged states, Ru(bpy)₃²⁺ and Ru(bpy)₃³⁺. The constructed gel exhibited a swelling transition when temperature was increased and the critical temperature of this phase transition depended on the state of Ru(bpy)$_3$ (Fig. 11.17). The gel swelled in the oxidized state because hydrophilicity of polymer chains was then increased. Hence, if temperature was kept fixed, oxidation of this atomic group could induce swelling, followed by deswelling when the reduction occurred.

When chemical oscillations took place, they were therefore accompanied by periodic swelling-deswelling phase transitions in the gel. A small bead of gel exhibited periodic contraction and expansion, similar to the heartbeat [30]. In such oscillations with the period of about 5 min, the chemical energy was transduced into the mechanical energy of the elastic gel. A more complex behavior was observed when rectangular slabs of the gel were used (Fig. 11.18). Chemical waves propagated along a slab and its length was periodically changing with time with the amplitude of about 50 $\mu$m. Investigations showed that the physical state of the gel changed the conditions for propagation of the waves, so that a feedback from the phase transition to the chemical reaction was involved [31]. Based on such chemomechanical phenomena, smart materials with applications to autonomous transport systems could be subsequently designed [32]. The theory of polymer gels with the BZ reaction was developed and reproduced well the observed effects [33].

Gels play a fundamental role in biological cells. The interior of an eukaryotic cell is occupied by cytoskeleton that represents a gel formed by cross-linked microtubules and actin filaments (Fig. 11.19). This gel is active, but different in its properties from the hydrogels coupled to nonequilibrium chemical reactions which we have just discussed. A single filament is a polar polymer chain made by actin monomers. It steadily grows at one end, where actin protein monomers from the solution become attached. The filament contracts at the opposite end, where actin molecules are detached. The arriving actin monomers have ATP bound to it. Inside the filament, the hydrolysis

**Fig. 11.17** Swelling ratio of the gel as a function of temperature for the oxidized and reduced states of the catalytic Ru group. Reproduced with permission from [30]

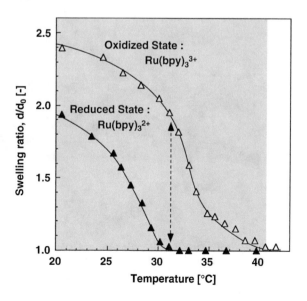

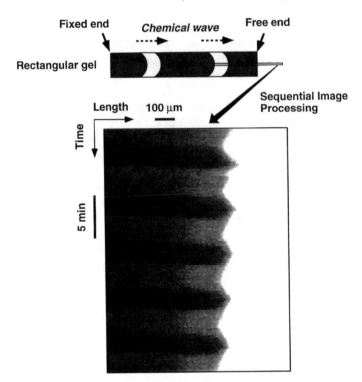

**Fig. 11.18** Oscillations in a thin gel slab with the BZ reaction. Propagation of the chemical wave induces periodic contractions of the gel. Reproduced with permission from [30]

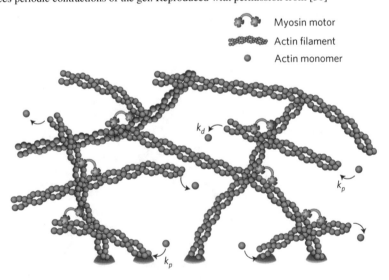

**Fig. 11.19** Active gel formed by actin filaments connected by myosin motors. Additional passive cross-links between the filaments are not displayed. Reproduced with permission from [34]

reaction takes place and ATP is converted into ADP, so that, when the monomers detach, they have ADP bound to them. Thus, already a single actin filament represents a nonequilibrium, i.e. *active*, molecular structure. A similar behavior is characteristic for microtubules formed by the protein tubulin. Moreover, in addition to passive, there are also active links that connect linear filaments into a three-dimensional gel. The active links are made by molecular motors, i.e. by the protein myosin. The myosin molecules, also converting ATP into ADP, can connect to different filaments and walk on them. As a result, a soft matter system far from thermal equilibrium is obtained. In contrast to classical passive gels that tend to approach an equilibrium stationary state determined by the minimum of their free energy, such active gels are capable of complex autonomous dynamics. Theoretical approaches to the description of biological active gels were reviewed by J. Prost, F. Jülicher and J.-F. Joanny [34].

# References

1. J.W. Cahn, J.E. Hilliard, J. Chem. Phys. **28**, 258 (1958)
2. S.C. Glotzer, E.A. Di Marzio, M. Muthukumar, Phys. Rev. Lett. **74**, 2034 (1995)
3. Q. Tran-Cong, A. Harada, Phys. Rev. Lett. **76**, 1162 (1996)
4. Q. Tran-Cong, J. Kawai, K. Endoh, Chaos **9**, 298 (1999)
5. M. Hildebrand, A.S. Mikhailov, G. Ertl, Phys. Rev. E **58**, 5483 (1998)
6. M. Hildebrand, A.S. Mikhailov, J. Phys. Chem. **100**, 19089 (1996)
7. M. Hildebrand, A.S. Mikhailov, G. Ertl, Phys. Rev. Lett. **81**, 2602 (1998)
8. Y. De Decker, H. Marbach, M. Hinz, S. Günther, M. Kiskinova, A.S. Mikhailov, R. Imbihl, Phys. Rev. Lett. **92**, 198305 (2004)
9. A. Locatelli, T.O. Mentes, L. Aballe, A.S. Mikhailov, M. Kiskinova, J. Phys. Chem. A **110**, 19108 (2006)
10. Y. De Decker, A.S. Mikhailov, J. Phys. Chem. B **108**, 14759 (2004)
11. M. Hildebrand, M. Kuperman, H. Wio, A.S. Mikhailov, G. Ertl, Phys. Rev. Lett. **83**, 1475 (1999)
12. M. Hildebrand, M. Ipsen, A.S. Mikhailov, G. Ertl, New J. Phys. **5**, 61 (2003)
13. A.S. Mikhailov, G. Ertl, Chem. Phys. Chem. **10**, 86 (2009)
14. A.S. Mikhailov, G. Ertl, Science **272**, 1596 (1996)
15. M. Seul, D. Andelman, Science **267**, 476 (1997)
16. M. Motoyama, T. Ohta, J. Phys. Soc. Jpn. **66**, 2715 (1997)
17. Y. Tabe, H. Yokoyama, Langmuir **11**, 4609 (1995)
18. R. Reigada, A.S. Mikhailov, F. Sagues, Phys. Rev. E **69**, 041103 (2004)
19. R. Reigada, F. Sagues, A.S. Mikhailov, Phys. Rev. Lett. **89**, 038301 (2002)
20. T. Okuzono, Y. Tabe, H. Yokoyama, Phys. Rev. E **69**, 050701 (2004)
21. J. Ignés-Mullol, J. Claret, R. Reigada, F. Sagues, Phys. Rep. **448**, 163 (2007)
22. Y. Tabe, H. Yokoyama, Nat. Mater. **2**, 806 (2003)
23. Y. Tsori, P.-G. de Gennes, Eur. Phys. J. E **14**, 91 (2004)
24. Y. Tabe, private communication
25. T. Shibata, A.S. Mikhailov, Europhys. Lett. **73**, 436 (2006)
26. M.-J. Huang, R. Kapral, A.S. Mikhailov, H.-Y. Chen, J. Chem. Phys. **137**, 055101 (2012)
27. J. Prost, R. Bruinsma, Europhys. Lett. **33**, 321 (1996)
28. J.-B. Manneville, P. Bassereau, D. Lévy, J. Prost, Phys. Rev. Lett. **82**, 4356 (1999)
29. M. Loose, E. Fischer-Friedrich, J. Ries, K. Kruse, P. Schwille, Science **320**, 789 (2008)
30. R. Yoshida, E. Kokufata, T. Yamaguchi, Chaos **9**, 260 (1999)

31. K. Miyakawa, F. Sakamoto, R. Yoshida, E. Kokufata, T. Yamaguchi, Phys. Rev. E **62**, 793 (2000)
32. R. Yoshida, Self-oscillating polymer gels as smart materials, in *Engineering of Chemical Complexity*, ed. by A.S. Mikhailov, G. Ertl (World Scientific, Singapore, 2013), pp. 169–185
33. V. Yashin, A.C. Balazc, Science **314**, 798 (2006)
34. J. Prost, F. Jülicher, J.-F. Joanny, Nat. Phys. **11**, 111 (2015)

# Chapter 12
# Molecular Machines

Fundamental functions in a living cell are performed at the level of single molecules. Some proteins act as enzymes and catalyze chemical reactions, while other operate as motors that generate mechanical forces and transport particles along filaments inside a cell. The proteins can act as pumps maintaining gradients of ion concentrations across a cellular membrane. There are also various proteins that perform operations with RNA and DNA, cutting them into pieces or glueing together. As the molecular biologist Bruce Alberts remarked [1] "the entire cell can be viewed as a factory that contains an elaborate network of interlocking assembly lines, each of which is composed of a set of large protein machines... Like the machines invented by humans to deal efficiently with the macroscopic world, these protein assemblies contain highly coordinated parts". Hence, Schrödinger's question *What is life?* cannot be fully answered without considering molecular machines.

Myosin is a molecular motor that provides a paradigmatic example of a protein machine and is responsible for force generation in the muscle. It was isolated in 1864 by W. Kühne from muscle extracts [2]. As shown in 1939 by V. Engelhardt and M. Lyubimova [3], myosin is an enzyme that catalyzes the hydrolysis of adenosine triphosphate (ATP), i.e. a reaction $ATP \rightarrow ADP + Pi$ by which ATP it converted to adenosine diphosphate (ADP) and the phosphate (Pi) is also produced. It was further discovered [4] by Albert Szent-Györgyi (who received a Nobel prize in medicine in 1937 for studies of vitamin C) that myosin can induce contraction only if another protein, actin, is present. As Szent-Györgyi recollected [5]:

> I repeated what W. Kühne did a hundred years earlier. I extracted myosin with strong potassium chloride (KCl) and kept my eyes open. With my associate, I. Banga, we observed that if the extraction was prolonged, a more sticky extract was obtained without extracting much more protein. We soon found that this change was due to the appearance of a new protein "actin", isolated in a very elegant piece of work by my pupil, F. Straub, while I "crystallized" myosin. Myosin, evidently, was a contractile protein, but the trouble was that *in vitro* it would do nothing.... So we made threads of the highly viscous new complex of actin and myosin, "actomyosin", and added boiled muscle juice. The threads contracted. To see them contract

© Springer International Publishing AG 2017
A.S. Mikhailov and G. Ertl, *Chemical Complexity*, The Frontiers Collection,
DOI 10.1007/978-3-319-57377-9_12

for the first time, and to have reproduced *in vitro* one of the oldest signs of life, motion, was perhaps the most thrilling moment of my life. A little cookery soon showed that what made it contract was ATP and ions.

The structure of the muscle was analyzed in 1953 by Hugh Huxley by means of electron microscopy [6]. It was found that both myosin and actin form long filaments that are arranged into a regular array, with myosin "thick" filaments placed between the "thin" filaments formed by actin. Based on his studies, Huxley has proposed [6] that stretching of a muscle takes place not by extension of the filaments, but by a process in which two sets of filaments slide past each other. Such sliding filaments mechanism was elucidated by using optical microscopy methods in the articles by Andrew Huxley (who has later got a Nobel prize for the discovery of ion mechanisms of nerve action) and Rolf Niedergerke [7] and by Hugh Huxley and Jean Hanson [8] that have both appeared in 1954 in the same issue of *Nature*.

Because the biological muscle is a complex object, in vitro experiments with reconstituted systems were performed in order to investigate interactions between myosin and actin. M. Sheetz and J. Spudich have adsorbed myosin molecules on the surface of fluorescent beads with the diameter of 0.7 $\mu$m and put such beads onto an array of actin cable from the *Nitella* alga. When ATP was supplied, directed motion of the beads could be observed [9]. Later on, a reverse motility assay was developed [10] where myosin filaments were immobilized on a glass surface and actin filaments were free to move over it. Using this setup with myosin molecules, Y. Harada et al. [11] demonstrated that they induce motion of actin filaments in the presence of ATP. Figure 12.1 shows the observed active motion of short fluorescent filaments. To determine trajectories, snapshots separated by 1.5 s were superposed.

In his theory of muscle function, Hugh Huxley has proposed [12] in 1969 that myosin heads bind to actin filaments (forming a "cross-bridge"), pull them and then detach. To check the theory and to study how myosin motors generate forces, experiments with motility assays had to be extended to the level of single molecules. Some complications however arise when working with muscle myosin (also known as myosin II). In a solution, its molecules tend to assemble themselves into thick filaments. Moreover, myosin II dissociates completely from the actin filament after completing an ATPase cycle.

There is a family of myosin motors that includes about 17 different proteins that have different functions. Myosin V is a motor that is employed to transport cargo along actin filaments to various destinations within the cell. It is a dimer with two actin-binding heads and two connected lever arms. If added to a solution with ATP, it gets attached to an actin filament and moves along it, dissociating after about 40–60 cycles. Hence, while being an important molecular motor, myosin V is also a convenient experimental object and many single-molecule investigations were performed using it.

Various advanced techniques were applied (and, sometimes, first developed) to monitor and control the motion of myosin V along the actin filament. In the experiments by M. Rief et al. [13], polystyrene 1 $\mu$m beads, sparsely coated with myosin V, were optically trapped in a focused laser beam and positioned near a surface-

**Fig. 12.1** ATP-induced
active motion of actin
filaments over the surface
coated by single-headed
myosin. *Dotted lines*
trajectories of fluorescent
filaments. *Scale bar* 5 μm.
Reproduced with permission
from [11]

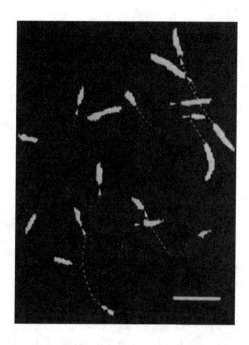

immobilized actin filament (Fig. 12.2). The motion of the beads could by optically
observed. In this setup, if the trap were stationary, it would have pulled back the
bead and stopped its motion after a few steps. To prevent this, a feedback scheme
was employed. The position of the digitally controlled optical trap was persistently
adjusted to the latest recorded bead position, so that the trap was moved after the
bead at a fixed separation from it.

Examining the time dependence in Fig. 12.2, one can notice that the motion of
myosin consists of quick steps by about 36 nm that are separated by the dwell time
intervals of about 0.1 s during which the motor does not move. It was demonstrated
that the dwell times were determined by the release of the hydrolysis product, ADP.

Based on their experimental data, the authors have proposed [13] the hand-over-
hand, or walking, mode for the processive motion of myosin V. To explain it, imagine
that the myosin dimer has two legs (representing levers with the head domains at
their ends). In each step, one of the legs remains attached to the filament, while the
other is moved over to the next position along it. The walking mode of myosin V was
subsequently confirmed [14] in the experiments where the technique of fluorescence
imaging with 1-nm accuracy was introduced.

Under fluorescence microscopy, positions of single protein molecules can be pre-
cisely tracked over time, but the molecules are not seen. What is then observed is the
dynamic behavior of individual fluorescent spots, not of the proteins themselves. To
study conformational changes, other methods, such as X-ray diffraction protein crys-
tallography, electron microscopy and nuclear magnetic resonance (NMR), are used.
However, they provide only static snapshots. For a long time, it seemed unfeasible
to see single biomolecules in action, with their conformational motions resolved.

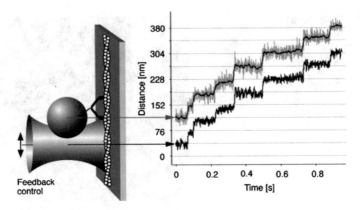

**Fig. 12.2** The experimental setup (*left*) and an example of the data (*right*) for optical tracking of myosin V along an actin filament. Via a digital feedback, the position of the laser trap (*black curve*) was persistently adjusted to keep a fixed separation from the current position of the bead (*grey curve*). Reproduced with permission from [13]

Nonetheless, such concomitant observations became possible due to a progress in atomic force microscopy (AFM).

In 2010, direct video imaging of walking myosin V by means of high-speed AFM was reported by N. Kodera et al. [15]. In their experiments, actin filaments were immobilized on lipid bilayers on a silicate surface. Tail-truncated myosin V molecules were put into the water solution and ATP was also supplied. Figure 12.3a shows a sequence of images where a single myosin molecule moves forward along the actin filament (the full video is available as Supplementary Information to Ref. [15]). The system is also schematically drawn in Fig. 12.3b. The directed motion consisted of discrete steps of 36 nm length and the measured average translocation velocity followed the Michaelis–Menten dependence on the concentration of ATP.

Because each advance by 36 nm in Fig. 12.3a was taking place within one video frame (146.7 ms), the molecular process during a step could not then be resolved. However, when streptavidin molecules acting as obstacles were added to the solution, they slowed the swinging arm motion and allowed visualization of the process. As seen in Fig. 12.3d, the nearly straight leading neck swings changing its arrowhead orientation (cf. frame 2 in Fig. 12.3c). The detached T-head rotationally diffused around the advancing head (but no translational diffusion on actin occurred). Then, it became bound to a forward site on the actin filament, completing one step. Thus, the hand-over-hand movement was directly confirmed.

Single-molecule experiments with other molecular motors were also performed. In addition to actin filaments, the cell includes a network of microtubules made up by a different structural protein, tubulin. Kinesin motors are employed by the cell to transport (big) cargo over microtubules. In 1996, R. Vale et al. [16] have used low-background total internal reflection microscopy to visualize processive motion of individual fluorescently labelled kinesin molecules along a microtubule.

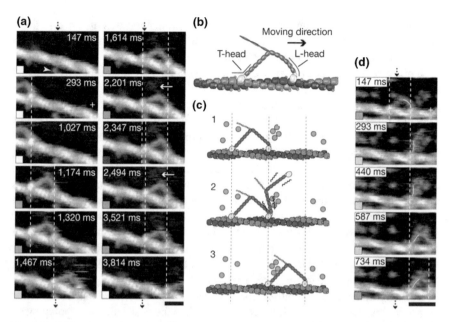

**Fig. 12.3** AFM observations of myosin V in action. **a** Successive AFM images showing processive motion of myosin. Scan area: $130 \times 65$ nm$^2$, *scale bar* 30 nm. **b, c** Schematics of the two-headed bound myosin and of its motion. *Small green dots* indicate streptavidin molecules added to slow the process. **d** AFM images showing hand-over-hand movement. The swinging lever is highlighted with a *thin light line*. Scan area: $150 \times 75$ nm$^2$, *scale bar* 50 nm. All images taken at 146.7 ms per frame. Reproduced with permission from [15]

**Fig. 12.4** Rotation of an actin filament attached to F1-ATPase. Adapted from [17]

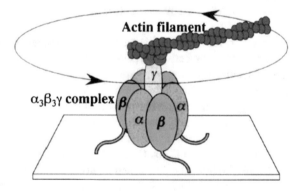

F1-ATPase is a membrane protein that consists of a ring formed by three $\alpha$ and three $\beta$ subunits, and of the central subunit $\gamma$ (Fig. 12.4). Each $\beta$ subunit has an active center and can catalyze the hydrolysis or synthesis of ATP. In the cell, this molecule is used to synthesize ATP, but it can also operate as a rotary motor when ATP is supplied. In 1997, Paul Boyer and John Walker received a Nobel prize for elucidation of the chemical mechanism of this molecular machine. The motor operation of F1-ATPase was demonstrated by H. Noji et al. [17]. As shown in Fig. 12.4, the molecule was fixed on a surface and a fluorescently labelled actin filament was attached to the $\gamma$ on top of it. By means of an epifluorescence microscope, steady rotational motion of the actin filament in presence of ATP was observed.

While motor proteins were most extensively studied, they represent only a special kind of molecular machines. Typically, the function of a machine is not to generate mechanical work, but to perform operations with other molecules in the cell. As an example, one can take the hepatitis C virus (HCV) helicase. Powered by ATP, this protein can unzip duplex DNA into two single strands. Experiments suggest [18, 19] that motor activity is involved in its operation. The protein consists of three domains, with two of them actively translocating along the upper DNA strand. The third domain is dragged by them and acts as a pulled wedge that mechanically separates the strands. Instead of the hand-over-hand movements characteristic for myosins, the inchworm mechanism is operating. Within each cycle, the left domain is detached from the DNA strand and moves closer to the immobile right domain. After that, it grasps the strand, whereas the right domain becomes dissociated from it and moves a step forward along the DNA. Finally, it grasps the DNA, whereas the left domain loosens its grip, and the cycle is repeated again.

Even highly complex operations can be carried out by individual machines. When human immunodeficit virus (HIV) replicates itself in the host cell, a long polypeptide chain is first formed. Different parts of this chain correspond to different proteins. The chain needs to be cut at precise positions to yield fragments representing various proteins, from which a new copy of the virus is later assembled. The cutting, or cleavage, is performed by a special molecular machine, HIV protease. This protein moves over the polypeptide chain, detects the desired positions, and behaves like molecular scissors cutting the chain. Twelve such operations have to be performed in a strict order by the machine. Remarkably, ATP or other additional substrates are not required for this.

In a fact, HIV protease is an enzyme characterized by a complex, but ordered, pattern of catalytic activity. While typically an enzyme catalyzes only one reaction, this molecular machine can implement a sequence of many different reactions in a coordinated way. There are also many other chemical nanomachines in the cell. As an example, below we consider the channeling enzyme tryptophan synthase.

While higher organisms take up the amino acid tryptophan with their food, bacteria, plants and yeast synthesize it. The synthesis is complicated and consists of 13 different reaction steps. One of the substrates is scarce in the cell and therefore the synthesis should be performed in an economical way. Moreover, an intermediate product, indole, is hydrophobic and can easily escape through the cell membrane; therefore, it should not be released into the cytoplasm. To satisfy such limitations, nature has found an elegant solution. The entire synthesis is performed inside a single biomolecule, tryptophan synthase, that has two subunits ($\alpha$ and $\beta$) each catalyzing a sequence of chemical reaction steps. The intermediate indole is never released into the solution, but channeled directly from one subunit to the other through a tunnel inside the enzyme.

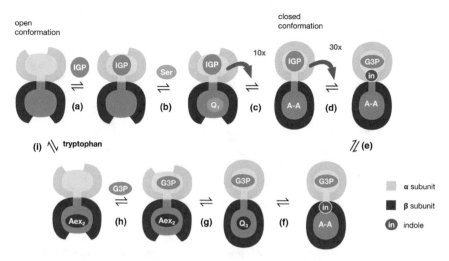

**Fig. 12.5** The catalytic cycle of tryptophan synthase. $Aex_2$ denotes the state of the $\beta$ subunit with tryptophan inside it. Reproduced from [20]

The simplified catalytic cycle of tryptophan synthase is shown in Fig. 12.5. Soon after the substrates have arrived to both subunits, their gates become closed (Fig. 12.5c) and the reaction proceeds in isolation within the enzyme. Indole is formed inside the $\alpha$ subunit (Fig. 12.5d) and is channeled from it to the $\beta$ subunit (Fig. 12.5f). When tryptophan and another product are synthesized, the gates become open (Fig. 12.5g) and the products leave. An extensive pattern of cross-regulation between the subunits is characteristic for this enzyme. The opening and closing of the gates is controlled by the presence of ligands in both subunits and, moreover, the reactions in one subunit can be activated or inhibited depending on the chemical state of the other one. As a result, synchronization between the reaction processes in the subunits takes place. In the reviews devoted to tryptophan synthase, it is described as "an allosteric molecular factory" [21] and "a channeling nanomachine" [22]. The enzyme has a domain structure and, as evidenced by X-ray diffraction experiments, the regulation is based on conformational transitions changing the shapes and relative positions of the domains.

Functional internal motions within macromolecules are the common property of various biological molecular machines. In each cycle, such ordered motions are robustly repeated again and again so that the analogy with macroscopic man-made mechanical machines is justified. In protein motors, such cyclic conformational changes are converted using a ratchet mechanism to steady translational or rotational motions, thus generating a force. Their function can however also be to facilitate chemical reactions or to perform operations with other biomolecules inside the cell.

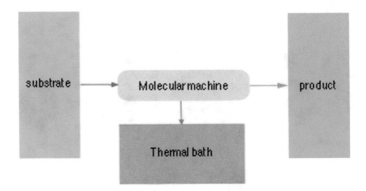

**Fig. 12.6** Molecular machine as an open thermodynamic system

The ordered internal dynamics of molecular machines is in a sharp contrast with random thermal fluctuations characteristic for an equilibrium state. What is the origin of such highly organized and coordinated microscopic behavior?

It should be noted that all biological protein machines represent enzymes and thus are catalytically converting a substrate into a product, as shown schematically in Fig. 12.6. In the motors, the substrate often represents ATP and its hydrolysis is catalyzed by them. Various other substrates can however be involved too, as in the above examples of HIV protease or tryptophan synthase. Typically, the substrate and product concentrations are maintained constant, so that there are two *chemostats*. Additionally, the molecular machine is always coupled to the thermal bath. If chemical potentials of the substrate and the product are different, this open molecular system will generally be in a non-equilibrium state.

In a fact, we have already considered such open macroscopic chemical systems that represent flow reactors (Fig. 2.3 in Chap. 2). The difference is that, in the case of molecular machines, an open system consists of a *single molecule* placed between two different chemostats. We have also seen in the previous chapters that various self-organization processes can proceed in open thermodynamic systems, leading to the development of concentration patterns and material structures on macro- and mesoscales. For a molecular machine, self-organization occurs within the macromolecule itself.

While being very small, molecules can still consist of definite parts, or domains. If a molecule is under non-equilibrium conditions, as in Fig. 12.6, motions of the domains do not represent thermal fluctuations. They may be well coordinated and strongly organized. Essentially, we deal then with self-organization processes at the molecular level.

Self-organized intramolecular motions can affect chemical reactions, regulating the catalytic activity of an enzyme. They can also influence interactions with other molecules, so that various operations with these molecules are performed. Eventually, because such motions are nonthermal, mechanical work can be produced and the molecule can function as a motor.

In Chap. 2 we have shown how thermodynamics can be applied to macroscopic open systems, leading to general predictions and constraints. These results cannot be directly transferred to microscopic systems because fluctuations in their variables are large. Nonetheless, as explained later in this chapter, *stochastic thermodynamics* can still be developed for such cases.

We have also seen in the previous chapters that, while the thermodynamic theory provides valuable general results, it does not itself tell what kinds of self-organization processes are possible in a particular system and how to describe them. In order to do this, specific models of reaction-diffusion phenomena or of soft matter systems had to be explored. The situation is similar for molecular machines: While the origins of their functional operation are clear, specific mechanisms involved in this operation need to be investigated and understood.

For a machine to operate, it should consist of rigid parts connected by joints. Moreover, while the joints have to be flexible, they should allow only ordered motions compliant with the function of the device. Industrial machines are typically made of metal parts with flexible connections between them. In the cell, soft matter is used instead. Biological machines represent proteins that are made by long peptide (or amino acid) chains.

The characteristic property of proteins is that they fold into a compact conformation which is uniquely defined (and therefore their folding is a self-assembly process). Thus, they agree well with the idea of an "aperiodic crystal" formulated by Schrödinger, though he originally introduced this concept referring to the molecular basis of genetic information transfer. In their compact form, proteins are stiff. However, proteins can also have a domain structure where stiff parts are connected by flexible junctions, such as a hinge. Therefore, they are suitable to build molecular machines.

Proteins typically consist of hundreds of amino acids, or residues, and their sizes in a folded state are in the nanometer range. They are functional in water solutions and their intramolecular relative domain motions are on the milliseconds scale. Equilibrium conformations of proteins (sometimes with various ligands) are known with atomic resolution for most proteins. These data are determined by crystallizing proteins and using the X-ray diffraction analysis. However, there are no experimental methods that would have allowed to follow conformational dynamics with a fine resolution and, therefore, computer simulations have to be performed.

All-atom molecular dynamics (MD) simulations for proteins can be readily undertaken, but the problem is that they take too much computer time. Unless for very small proteins, current simulations cover only microseconds, which is much shorter than the millisecond times typical for functional conformational motions in protein machines. Therefore, approximate coarse-grained descriptions have to applied.

In the approach known as *elastic network (EN) modeling*, a protein is modeled as a network of particles connected by elastic springs [23, 24]. Thus, its energy is given by

$$E = \frac{k}{4} \sum_{i,j=1}^{N} A_{ij} \left( d_{ij} - d_{ij}^0 \right)^2 \tag{12.1}$$

**Fig. 12.7** The elastic
network of myosin V. The
actin binding cleft (*left lower
corner*) and the nucleotide
binding pocket are
highlighted by *light grey*.
Reproduced from [26]

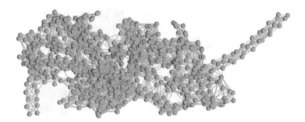

where $d_{ij}$ is the length of a spring connecting particles $i$ and $j$ and $d_{ij}^0$ is the natural length of this spring.

The EN models are structure-based: the pattern of connections is determined by the experimentally known equilibrium spatial structure of a protein. The equilibrium positions $R_i^0$ of all residues are taken from the X-ray diffraction data and equilibrium distances $d_{ij}^0 = \left| R_i^0 - R_j^0 \right|$ are thus found. If the equilibrium distance between particles $i$ and $j$ is shorter than the chosen cut-off length, the two particles are connected by a spring and $A_{ij} = 1$. Otherwise, the link is absent and $A_{ij} = 0$. In the simplest variant of the EN method, all springs are assumed to exhibit the same stiffness $k$. Spring constants can however also be dependent on the kinds of the residues and on the equilibrium distances between them [25].

As an example, Fig. 12.7 shows the elastic network of the head of myosin V. The network consists of $N = 752$ beads and a large number of links. A short fragment of the arm is also included into it. The ATP arrives from the solution into the pocket located on the top; the myosin holds the actin filament in the cleft seen in the lower bottom part.

On the time scales characteristic for slow conformational motions in solutions, inertial effects are negligible and the motion of beads is described by equations

$$\frac{d R_i}{dt} = -\gamma \frac{\partial E}{\partial R_i} \tag{12.2}$$

for their time-dependent positions $R_i(t)$. Here, $\gamma$ is the mobility coefficient of the beads. Fluctuations can be taken into account by adding thermal noise.

Together, Eqs. (12.1) and (12.2) describe the process of conformational relaxation starting from a deformed state of the network. Note that only relatively weak local deformations are allowed in the EN approach; it is not applicable to describe folding or unfolding of a protein. These dynamical equations are nonlinear because the distance between two particles

$$d_{ij} = \left| R_i - R_j \right| = \sqrt{\left( X_i - X_j \right)^2 + (Y_i - Y_j)^2 + (Z_i - Z_j)^2} \tag{12.3}$$

is a nonlinear function of their coordinates. By its construction, the network always has its absolute energy minimum $E = 0$ at the experimentally known equilibrium

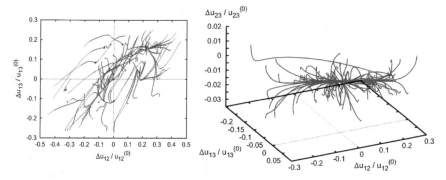

**Fig. 12.8** Relaxation trajectories in a random elastic network (*left*) and in the elastic network corresponding to the $\beta$-subunit of F1-ATPase (*right*). Final points of the trajectories are indicated by *blue dots*. Projections of the trajectories into the space of relative distances between three selected labels are displayed. The equilibrium state with zero energy lies in the origin of the coordinates. Reproduced from [27]

conformation of the protein. However, because of the nonlinearities, other energy minima, corresponding to metastable states, might also exist.

Figure 12.8 (left) shows 100 relaxation trajectories starting from different initial conditions in a *random* elastic network with 64 beads. The relaxation pattern is disordered and there are many metastable states. A completely different situation is encountered when a network corresponding to a machine protein is chosen. Figure 12.8 (right) displays 100 trajectories for a protein that represents a single $\beta$ subunit of the rotary motor F1-ATPase. Two labels (2 and 3) belong to the same stiff domain of the protein, while label 1 is located in a different mobile domain. In a clear contrast to the random network, all trajectories end in the same equilibrium state, even though the distance between the domains could have changed by up to 30%.

Examining Fig. 12.8 (right), a further important property of conformational relaxation in molecular protein machines can be noted. While the trajectories begin at different points, they soon converge to the same tight bundle that leads to the equilibrium state. The same behavior could be seen for other molecular motors, such as myosin and kinesin [28] and HCV helicase [29]. It implies that the elastic energy landscape of molecular machines is organized in a special manner: It has the *funnel structure* with a deep narrow valley that leads to the equilibrium state.

Thus, conformational relaxation in motor proteins proceeds in a highly ordered way. Although there are many degrees of freedom, a single attractive trajectory exists. This trajectory, corresponding to the bottom of a narrow energy valley, ends in the equilibrium state and the distance along it can be considered as the internal mechanical coordinate of the protein. It is as if the protein had a hidden rail track along which its domains could move. While the motion along the track can be easily induced and controled, large forces are needed to kick the protein off the track.

In Chap. 10, we have previously noticed that the existence of low-dimensional attractive manifolds, i.e. of the "hidden rail tracks", is a characteristic property of dynamics in self-organized systems. The collective variables, representing coordinates on such manifolds, play the role of order parameters of a system. Because of such tracks, complex self-organized systems are able to exhibit robust and predictable dynamics that can be easily controlled. Now, we see that even single macromolecules can possess such dynamical properties and they are essential for their operation as machines. This special organization of the energy landscape in the folded conformation in motor proteins is a result of selection in the process of biological evolution. Remarkably, artificial elastic networks with similar dynamical properties could be constructed by using computer evolution algorithms [27].

In macroscopic mechanical motors, energy can be continuously supplied. In contrast to this, energy can be provided only in discrete portions, i.e. in the chemical form with the ligands, in molecular machines. Hence, the operation of such a machine represents an alteration of sudden excitation processes followed by conformational relaxation. The excitations can result from binding or release of a ligand, or from a catalytic chemical reaction that changes the ligand state within a protein. In other words, ligand-induced chemomechanical motions are then involved.

If domain motions in a protein are ordered and proceed "on a track", the protein will also respond to perturbations, such as ligand binding, in an ordered way, i.e. with the domains moving a little along the track in response to it. Therefore, the form of the response is not highly sensitive to the details of a perturbation in protein machines. In coarse-grained EN models, the effect of a ligand, such as ATP, can be imitated by introducing an additional particle into an elastic network. This particle interacts with its neighbours and a conformational change is induced.

With such modeling, the entire operation cycle of the molecular motor HCV helicase could be reproduced in a structurally resolved way [29]. Figure 12.9 shows three consequent snapshots from a video illustrating the simulation results. The upper two domains (brown and blue) actively translocate along the upper DNA strand. In each cycle, a ligand particle (mimicking the ATP) arrives into a binding pocket on the surface of the upper left domain. It induces a mechanical motion bringing the two domains close one to another. When they touch, a reaction converting the ligand

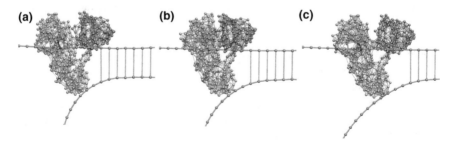

**Fig. 12.9**  Operation of HCV helicase. Three consequent snapshots from an EN simulation including the DNA. Two actively translocating motor domains (*brown* and *blue*) pull the third domain (*grey*) that acts as a wedge and unzips the two DNA strands (*green*). Reproduced from [29]

into a product takes place and the product is immediately released. After that, the domains move away one from another, recovering the initial conformational state.

The ligand-induced chemomechanical motion consists not only in bringing closer the two upper domains. Additionally, their interactions with DNA become modified. In the ligand-free open state, the right domain holds the DNA and the left domain is free to move. In contrast to this, in the closed state with the ligand the right domain ceases to hold the DNA strand and it is grasped by the left domain.

As a result of repeated ligand-induced conformational changes that affect interactions with the DNA, translocation of the two motor domains along the upper DNA strand takes place. The third domain is passively dragged and, acting as a wedge, separates the two DNA strands. Thus, the inchworm operation mechanism, that was proposed based on the experimental data [18, 19], has been theoretically confirmed.

It is surprising that simple mechanical models, treating proteins as elastic networks and neglecting many chemical details, can nonetheless well describe the operation of molecular machines. However, the agreement extends far beyond the qualitative level, as revealed by comparing the predictions of EN models with all-atom MD simulations for small proteins [30]. Apparently, elastic effective forces play a dominant role in slow conformational dynamics that controls functional chemomechanical motions of machine domains.

In coarse-grained dynamical models, conformational changes underlying the operation of a molecular machine were structurally and temporally resolved. There is however also a different description level where the machine operation is modelled as a sequence of stochastic transitions among a small set of internal states. In the above example of HCV helicase, there would be only two of them, i.e. the open and the closed conformations of the protein. When ligand binds, a transition from the open to the closed conformation takes place and a certain amount of work is produced. When the ligand is converted to the product and released, the reverse transition takes place and the protein returns to its initial state after converting one substrate into one product molecule.

Discrete stochastic descriptions are known as *Markov models* of molecular machines. To construct such a model, a set of distinct chemical states needs to be identified and rates of transitions between them have to be experimentally determined. After that, stochastic simulations can be performed. Markov descriptions are available, e.g., for molecular motors F1-ATPase [31, 32] and kinesin [33].

As an example, Fig. 12.10 shows the Markov transition network for the channeling enzyme tryptophan synthase (see also Fig. 12.5). Here, each node represents a combination of the possible states of the $\alpha$ and $\beta$ subunits, which are enumerated from 1 to 4 and from 1 to 6. The state of a subunit can change either because the respective ligands (substrates or products) bind or dissociate, or because of a chemical reaction event within it. The transition from (3, 3) to (4, 4) simultaneously changes the states of two subunits; it corresponds to channeling of the intermediate product (indole). All transitions are reversible; the experimentally measured reaction rate constants are indicated next to the arrows in Fig. 12.10. The mean turnover time of the enzyme is about 0.15 s.

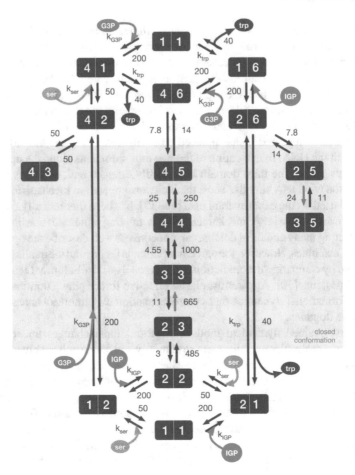

**Fig. 12.10** Markov transition network for a single molecule of the channeling enzyme tryptophan synthase. Reproduced from [20]

Although classical thermodynamics is not applicable for single molecules where fluctuations are strong, *stochastic thermodynamics* was developed to describe them [34–36]. Its starting point is the evolution equation (also known as the *master equation*) for probabilities $p_i$ to occupy different states $i$. Such probabilities can change with time because of the transitions between the states, leading to probability fluxes over a Markov network. Each state is characterized by its energy $E_i$ and, because the molecule is at equilibrium with a thermal bath at temperature $T$, conditions of the microscopic detailed balance, that connect the rates of transitions in the forward and reverse directions between any two states, are to be satisfied. The entropy of a microscopic system is defined in terms of the probabilities as $s = -\sum_i p_i \ln p_i$; it is measured in units of bits (1 bit = ln 2). By using the master equation, production and export rates of entropy can be determined; fluxes of entropy over a network can be found. Moreover, extensions of the First and the Second laws of classical thermodynamics become formulated.

For tryptophan synthase, Gibbs energies of all internal states could be identified by using the known reaction rates and the calorimetric data [20]. Under in vivo conditions, the difference of the Gibbs energy between the substrates and the products of this enzyme is $\Delta G = 19.56\,k_B T$ and thus it operates far from equilibrium, with this amount of energy dissipated in each cycle. The channeling transition is driven by the energy difference of $5.4\,k_B T$ and does not therefore represent a diffusion-like process. In every catalytic cycle, the molecule generates 27.79 bits of entropy and the same amount is exported to the environment around it. The production and export of entropy are nonuniformly distributed over the network.

Because of allosteric interactions between the subunits and of the channeling events, strong correlations between the states of the two subunits develop in the enzyme. The statistical correlation $i(a, b)$ between a state $a$ of the subunit $\alpha$ and a state $b$ of the subunit $\beta$ can be estimated as

$$i(a, b) = \ln \frac{p(a, b)}{p_\alpha(a) p_\beta(b)} \tag{12.4}$$

where $p(a, b)$ is the joint probability of having the two subunits in the states $a$ and $b$. Moreover, $p_\alpha(a) = \sum_{b=1}^{6} p(a, b)$ and $p_\beta(b) = \sum_{a=1}^{4} p(a, b)$ are the probabilities to have one of the subunits in the respective state. If the two states are independent, $p(a, b) = p(a) p(b)$ and $i(a, b) = 0$. The average $I_{\alpha\beta} = \sum_{a,b} p(a, b) i(a, b)$ gives the *mutual information* of the two subunits of the enzyme. For tryptophan synthase under physiological conditions, $I_{\alpha\beta} = 0.49$ bit.

Statistical correlations can also be viewed as specifying the degree of synchronization between the processes taking place within the two parts. It has been found that there is only weak correlation $i(1, 1) = 0.27$ between the empty states ($a = b = 1$) of the subunits. Strong correlations however develop along the main catalytic pathway and, in the state ($a = b = 3$) just before channeling, they reach $i(3, 3) = 1.6$.

This last example provides a good illustration of self-organization phenomena at the molecular level. Generally, chemical nanomachines behave like "factories" that perform ordered operations with single molecules in a self-organized way. To control the operations, they employ a system of gates. These gates can be used to block the access of substrates to catalytic sites or to release final products when they are formed. They can also enhance or inhibit the catalytic activity of an active center depending on the current states of other parts of the machine. Moreover, intermediate products can be passed from one active center to another within the machine without releasing them into the solution, as if there were a built-in conveyor belt. The gates of a chemical nanomachine can be considered as manipulated by *Maxwell demons*. A "demon" observes the molecular states in some parts and, based on such information, opens or closes the gates in its other parts.

From an abstract perspective, a chemical nanofactory can be seen as consisting of the superimposed material and information flow networks. In the material network, single molecules arrive as substrates from the solution and are processed (i.e., modified, assembled or cleaved) on a set of active operation sites. They are transported within the factory from one site to another and, at the end, the final products are released. The information about the occupation states of all active sites is gathered, processed and used to manipulate a set of molecular gates that control the material transport and the operations performed at the individual sites.

Remarkably, the information gathering and control of the gates have to be completely self-organized because, in contrast to an industrial factory, no external controlling agents (like humans or computers) can be employed. In proteins, the regulation is based on allosteric effects: binding of a ligand or a change in its chemical state at one protein site can induce conformational changes at another protein site, thus controlling the access and release of other ligands, or the catalytic activity, at this site.

In a living cell, further examples of chemical "nanofactories" can be found. As a growing amount of evidence suggests, many reactions within a cell may be catalyzed not by individual enzymes, but by multienzyme complexes (see, e.g., [37–42]). Such complexes consist of tens of different proteins and can implement entire metabolic pathways or significant parts of them. Within a complex, intermediate products can be directly channeled to other enzymes for further processing [38, 39] and, moreover, different enzymes in a complex are usually coupled through allosteric regulatory loops [42]. Experimental investigations of multienzyme complexes encounter difficulties because the complexes are often transient and only exist in vivo, inside a cell. There is however a class of channeling enzymes (see review [43] which are similar in their properties to multienzyme complexes, but, in contrast, they are smaller and stable. The considered above tryptophan synthase is one of such channeling enzymes.

To form enzyme complexes, special scaffold proteins are often used by the cell [44]. A scaffold provides a template onto which a number of other proteins can bind. If all enzymes corresponding to a certain complex reaction are attached to a scaffold, it becomes spatially localized. Additionally, the scaffold can mediate allosteric interactions between the attached enzymes, providing regulatory feedback loops. Based on the scaffolds, high-fidelity intracellular information transfer can be organized, with an an entire molecular signalling pathway implemented by the proteins on the same scaffold [44].

Molecular machines can have important technological applications outside of biological cells. For such applications, actual biological machines need to be modified or novel artificial machines should be designed and built. An example of a synthetic protein machine was provided by D. Hoersch, S.-H. Roh and T. Kortemme [45]. Chaperonins are protein assemblies that assist in folding of other proteins in the cell. Their operation is accompanied by large confirmational changes induced by the hydrolysis of ATP. Typically, a chaperonin is formed by a ring of linked protein subunits that form a cage into which a protein, that has to be fold, is put. Figure 12.11a shows the Mm chaperonin from *Methanococcus maripaludis* which resembles a barrel with a built-in lid. It has two eight-membered rings each enclosing a central chamber of about 150 nm$^3$ in size. The chamber is capped by a lid that opens and closes to encapsulate client substrate proteins in a cycle driven by ATP binding and hydrolysis.

Hoersch et al. [45] designed a mechanism that could switch between the closed and open states of this chaperonin by light (Fig. 12.11b). This was done by taking

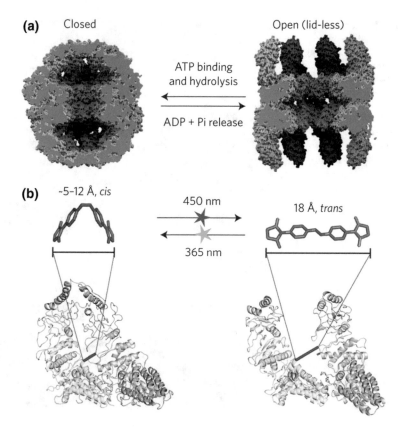

**Fig. 12.11** Mm chaperonin (**a**) and the design scheme (**b**) of its light-controlled synthetic version. Reproduced with permission from [45]

a small azobenzene-dimaleimide (ABDM) molecule and incorporating it inside the chaperonin. The ABDM molecule reversibly changes its end-to-end distance through *cis-trans* isomerization under illumination with either near ultraviolet or visible light. The photoinduced *cis-trans* transition induced in the chaperonin a conformational change that was analogous to the effect of ATP binding and hydrolysis. Thus, a light-operated nanocage was obtained. Figure 12.12 shows cryo-electron microscopy reconstructions of the unlit (left) and blue-light illuminated (right) conformations of the designed synthetic machine.

While protein machines, either natural or modified, are single molecules, they are still fairly large, typically comprising thousands of different atoms. Is it possible to build the machines on even shorter, atomic scales? This question was first addressed by Richard Feynmann in his talk [46] at the Annual Meeting of the American Physical Society in 1959. In 2016, Jean-Pierre Sauvage, James Fraser Stoddart and Bernard L. Feringa received the Nobel prize "for the design and synthesis of molecular machines".

To construct a machine, one needs molecular assemblies whose parts can relatively freely move with respect one to another but the structure does not break. One

**Fig. 12.12** The light-operated nanocage. Cryo-electron microscopy reconstructions of the unlit (*left*) and *blue-light* illuminated (*right*) conformations of the modified Mm chaperonin. Both the side and the top views are displayed. Reproduced with permission from [45]

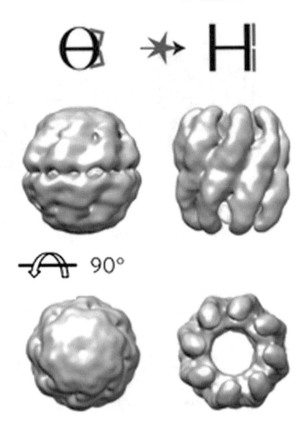

possibility to design such assemblies is to use topological entanglement, or mechanical bonds. In such molecules, the individual parts are not held together by covalent bonds, but are inseparably entangled. In catenanes (Fig. 12.13, top), two rings are interlocked. In rotaxanes (Fig. 12.13, bottom), a ring is threaded onto an axle and free to move along it, but cannot escape because of the stoppers at the axle ends. Both assemblies could be synthesized [47–49], but their synthesis was challenging and had an extremely low yield. In 1983, Sauvage with coworkers [50] introduced an efficient method of template synthesis that allowed to easily produce interlocked molecules of various kinds.

In 1991, Stoddart et al. produced a rotaxane that they described as the molecular shuttle [51]. Its structure is schematically illustrated in Fig. 12.14. The tetracationic "bead" (blue) is threaded on the polyether axle with two silyl stoppers (black and green) at the ends. The axle has two immobile "stations" (black and red) formed by hydroquinol rings. These stations can be addressed selectively by chemical, electrochemical, or photochemical means and provide a mechanism to drive the bead along the axle. As the authors noted, "insofar as it becomes possible to control the movement of one molecular component with respect to the other in a [2]rotaxane, the technology for building "molecular machines" will emerge".

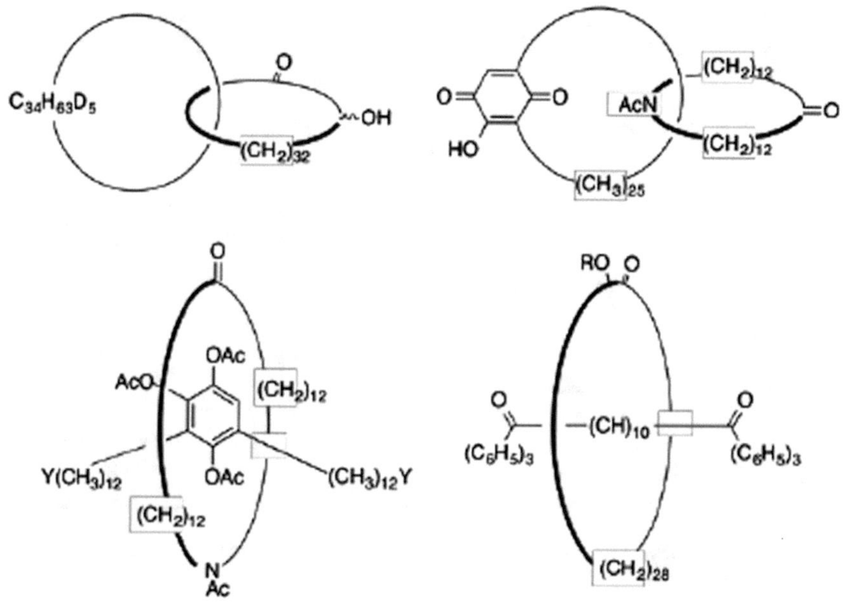

**Fig. 12.13** Examples of molecular structures based on mechanical bonds: catenanes (*top*) and rotaxanes (*bottom*)

Various molecular devices, based on rotaxane and catenane structures with topo-logical entanglement, were subsequently developed by Stoddart and Sauvage (see review [52]). As an example, we show in Fig. 12.15 the operation scheme of an artificial molecular motor powered by light [53].

The synthesized rotaxane $\mathbf{1}^{6+}$ molecule comprises a bis-*p*-phenylene-34-crown-10 electron donor macrocycle, shown as the ring in Fig. 12.15, and a dumbbell-shaped component that contains two electron acceptor recognition sites for the ring, namely a 4,4'-bipyridinium ($A_1^{2+}$) and a 3,3'-dimethyl-4,4'-bipyridinium ($A_2^{2+}$) units. These units play the role of "stations" for the ring and are colored blue and pink in Fig. 12.15. The molecule has the length about 5 nm and the distance between the centers of the two stations is 1.3 nm. The dumbbell-shaped component incorporates a $[Ru(bpy)_3]^{2+}$ - type electron transfer photosensitizer $P^{2+}$ (colored as green) that is able to operate with visible light and also plays the role of a stopper. Moreover, it includes a *p*-terphenyl-type rigid spacer S (yellow) that keeps the photosensitizer far from the electron acceptor units and a tetraarylmethane group **T** (gray) as the second stopper.

The operation cycle of this molecular machine consists of several stages (Fig. 12.15). First, light excitation $P^{2+} \rightarrow *P^{2+}$ of the photoactive unit takes place. It is followed by the transfer of an electron from the $*P^{2+}$ excited state to the $A_1^{2+}$ station, with the consequent "deactivation" of this station and the oxidation of the photosensitive unit to $P^{3+}$. After reduction of the $A_1^{2+}$ station to $A_1^+$, the ring moves

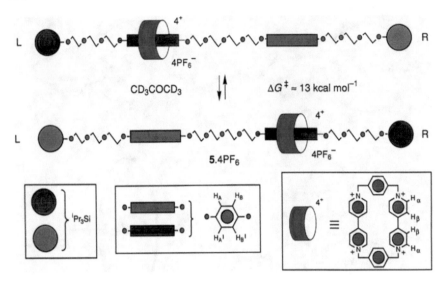

**Fig. 12.14** The scheme of a rotaxane operating as a molecular shuttle (see the text). Reproduced with permission from [51]

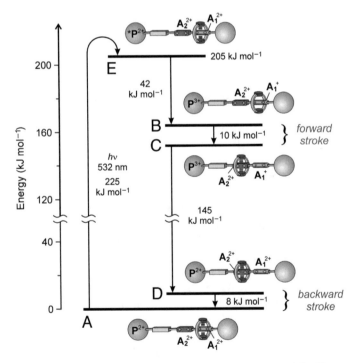

**Fig. 12.15** The operation mechanism of a molecular motor powered by light. The energy levels involved in the photoinduced ring shuttling at room temperature are indicated. Reproduced with permission from [53]

(the forward stroke) by Brownian motion to $A_2^{2+}$. Next, a back electron-transfer process from the "free" reduced station $A_1^+$ to $P^{3+}$ restores the electron acceptor power of the $A_1^{2+}$ station. As a consequence of the electronic reset, the ring moves back (the backward stroke) again by Brownian motion from $A_2^{2+}$ to $A_1^{2+}$, and the cycle becomes completed. The motor can be driven at a frequency of $10^3$ Hz, potentially generating a power of $3 \times 10^{-17}$ W per molecule.

A different route to the design of molecular machines, based on isomerization transitions, was chosen by Feringa. Above we have already seen that some molecules exist in two stable conformations, *trans* and *cis*, and that transitions between such conformations can be induced by light. N. Koumura et al. [54] synthesized a molecule where four discrete isomerization steps, activated by ultraviolet light or a temperature change, resulted in a full conformational rotation around a central carbon-carbon double bond. Thus, it behaved like a rotary motor. Such motors could be subsequently anchored on a gold surface [55] or incorporated into a liquid-crystal film and used to rotate micrometer-size objects [56].

By incorporating such artificial light-driven molecular motors into a polymer gel, N. Giuseppone with coworkers [57] could induce the unidirectional contraction of this material. Furthermore, reversible contraction-expansion cycles of the gel could be observed when two kinds of active elements, motors and modulators, were used [58]. Thus, a transition to macroscopic motions, similar to the operation of a biological muscle but employing different kinds of molecular machines, could be achieved.

These last examples demonstrate how single molecules, feeding on chemical energy or nurtured by light, can generate mechanical work without the need for biological systems, but according to the principles of self-organization as outlined in this book. Thus, the circle becomes closed now.

# References

1. B. Alberts, Cell **92**, 291 (1998)
2. W. Kühne, *Untersuchungen über das Protoplasma und die Contractilität* (W. Engelmann, Leipzig, 1864)
3. V.A. Engelhardt, M.N. Lyubimowa, Nature **144**, 668 (1939)
4. A. Szent-Györgyi, Biol. Bull. **96**, 140 (1949)
5. A. Szent-Györgyi, Annu. Rev. Biochem. **32**, 1 (1963)
6. H.E. Huxley, Biochim. Biophys. Acta **12**, 387 (1953)
7. A.F. Huxley, R. Niedergerke, Nature **173**, 971 (1954)
8. H.E. Huxley, J. Hanson, Nature **173**, 973 (1954)
9. M. Sheetz, J. Spudich, Nature **303**, 31 (1983)
10. S.J. Kron, J.A. Spudich, Proc. Natl Acad. Sci. U.S.A. **83**, 6272 (1986)
11. Y. Harada, A. Noguchi, A. Kishino, T. Yanagida, Nature **326**, 805 (1987)
12. H.E. Huxley, Science **164**, 1356 (1969)
13. M. Rief, R.S. Rock, A.D. Mehta, M.S. Mooseker, R.E. Cheney, J.A. Spudich, Proc. Natl Acad. Sci. U.S.A. **97**, 9482 (2000)
14. A. Yildiz, J.N. Forkey, S.A. McKinney, T. Ha, Y.L. Goldman, P.R. Selvin, Science **300**, 2061 (2003)

15.  N. Kodera, D. Yamamoto, R. Ishikawa, T. Ando, Nature **468**, 72 (2010)
16.  R.D. Vale, T. Funatsu, D.W. Pierce, L. Romberg, Y. Harada, T. Yanagida, Nature **380**, 451 (1996)
17.  H. Noji, R. Yasuda, M. Yoshida, K. Kinoshita Jr., Nature **386**, 299 (1997)
18.  S. Dumont, W. Cheng, V. Serebrov, R.K. Beran, I. Tinoco Jr., A.M. Pyle, C. Bustamante, Nature **439**, 105 (2006)
19.  S. Myong, M.C. Bruno, A.M. Pyle, T. Ha, Science **317**, 513 (2007)
20.  D. Loutchko, A.S. Mikhailov, J. Chem. Phys. **146**, 025101 (2017)
21.  T.R. Barends, M.F. Dunn, I. Schlichting, Curr. Opin. Chem. Biol. **12**, 593 (2008)
22.  M.F. Dunn, D. Niks, H. Ngo, T.R.M. Barends, I. Schlichting, Trends Biochem. Sci. **33**, 254 (2008)
23.  M. Tirion, Phys. Rev. Lett. **77**, 1905 (1996)
24.  T. Haliloglu, I. Bihar, B. Erman, Phys. Rev. Lett. **79**, 3090 (1997)
25.  Y. Dehouck, A.S. Mikhailov, PLoS Comput. Biol. **9**, e1003209 (2013)
26.  M. Düttmann, Y. Togashi, T. Yanagida, A.S. Mikhailov, Biophys. J. **102**, 542 (2012)
27.  Y. Togashi, A.S. Mikhailov, Proc. Natl Acad. Sci. U.S.A. **104**, 8697 (2007)
28.  Y. Togashi, T. Yanagida, A.S. Mikhailov, PLoS Comput. Biol. **6**, e1000814 (2010)
29.  H. Flechsig, A.S. Mikhailov, Proc. Natl. Acad. Sci. U.S.A. **107**, 20875 (2010)
30.  M. Gur, E. Zomot, I. Bihar, J. Chem. Phys. **139**, 121912 (2013)
31.  P. Gaspard, E. Gerritsma, J. Theor. Biol. **247**, 672 (2007)
32.  K. Kawaguchi, S.I. Sasa, T. Sagawa, Biophys. J. **106**, 2450 (2010)
33.  S. Liepelt, R. Lipowsky, Phys. Rev. **79**, 011917 (2009)
34.  J. Schnakenberg, Rev. Mod. Phys. **48**, 571 (1976)
35.  K. Sekimoto, J. Phys. Soc. Jpn. **66**, 1234 (1997)
36.  C. Jarzynski, Phys. Rev. Lett. **78**, 2690 (1997)
37.  Z.H. Zhou, D.B. McCarthy, C.M. O'Connor, L.J. Reed, J.K. Stoops, Proc. Natl Acad. Sci. U.S.A. **98**, 14802 (2001)
38.  C.Z. Constantine et al., Biochemistry **45**, 8193 (2006)
39.  S. Li et al., J. Mol. Biol. **366**, 1603 (2007)
40.  M.V. Górna, A.J. Carpousis, B.F. Lousi, Q. Rev. Biophys. **45**, 105 (2012)
41.  A.M. van Oijen, J.J. Loparo, Annu. Rev. Biophys. **39**, 429 (2010)
42.  R. Nussinov, B. Ma, C.J. Tsai, Biochim. Biophys. Acta **1834**, 820 (2012)
43.  X. Huang, H. Holden, F.M. Raushel, Annu. Rev. Biochem. **70**, 149 (2001)
44.  M.C. Good, J.G. Zalatan, W.A. Lim, Science **332**, 680 (2011)
45.  D. Hoersch, S.-H. Roh, T. Kortemme, Nat. Nanotechnol. **8**, 928 (2013)
46.  R. Feynmann, "There is plenty of room at the bottom" (Transcript of the talk given on December 29, 1959, at the Annual Meeting of the American Physical Society). Caltech Eng. Sci. **23**, 22 (1960)
47.  E. Wasserman, J. Am. Chem. Soc. **82**, 4433 (1960)
48.  G. Schill, A. Lüttringhaus, Angew. Chem. Int. Ed. **3**, 546 (1964)
49.  I.T. Harrison, S. Harrison, J. Am. Chem. Soc. **89**, 5723 (1967)
50.  C.O. Dietrich-Buchecker, J.P. Sauvage, J.P. Kintzinger, Tetrahedron Lett. **24**, 5095–5098 (1983)
51.  P.L. Aneli, N. Spenser, J.F. Stoddart, J. Am. Chem. Soc. **113**, 5131 (1991)
52.  V. Balzani, A. Credi, F.R. Raymo, J.F. Stoddart, Angew. Chem. Int. Ed. **39**, 3348 (2000)
53.  V. Balzani, M. Clemente-León, A. Credi, B. Ferrer, M. Venturi, A.H. Flood, J.F. Stoddart, Proc. Natl Acad. Sci. U.S.A. **103**, 1178 (2006)
54.  N. Koumura, R.W. Zijstra, R.A. van Delden, N. Harada, B.L. Feringa, Nature **401**, 152 (1999)
55.  R.A. van Delden, M.K.J. ter Wiel, M.M. Polard, J. Vicario, N. Koumura, B.L. Feringa, Nature **437**, 1337 (2005)
56.  R. Eelkema, M.M. Polard, J. Vicario, N. Katsonis, B.S. Ramon, C.W.M. Bastiaansen, D.J. Broer, B.L. Feringa, Nature **440**, 163 (2006)
57.  Q. Li, G. Fuks, E. Moulin, M. Rawiso, I. Kulic, J.T. Foy, N. Giuseppone, Nat. Nanotechnology **10**, 161 (2015)
58.  J.T. Foy, Q. Li, A. Goujon, J.-R. Colard-Itte, G. Fuks, E. Moulin, O. Schliffmann, D. Dattler, D.P. Funeriu, N, Giuseppone, Nat. Nanotechnology **12**, 540 (2017)

# Titles in this Series

**Quantum Mechanics and Gravity**
By Mendel Sachs

**Quantum-Classical Correspondence**
Dynamical Quantization and the Classical Limit
By A.O. Bolivar

**Knowledge and the World: Challenges Beyond the Science Wars**
Ed. by M. Carrier, J. Roggenhofer, G. Küppers and P. Blanchard

**Quantum-Classical Analogies**
By Daniela Dragoman and Mircea Dragoman

**Quo Vadis Quantum Mechanics?**
Ed. by Avshalom C. Elitzur, Shahar Dolev and Nancy Kolenda

**Information and Its Role in Nature**
By Juan G. Roederer

**Extreme Events in Nature and Society**
Ed. by Sergio Albeverio, Volker Jentsch and Holger Kantz

**The Thermodynamic Machinery of Life**
By Michal Kurzynski

**Weak Links**
The Universal Key to the Stability of Networks and Complex Systems
By Csermely Peter

**The Emerging Physics of Consciousness**
Ed. by Jack A. Tuszynski

**Quantum Mechanics at the Crossroads**
New Perspectives from History, Philosophy and Physics
Ed. by James Evans and Alan S. Thorndike

**Mind, Matter and the Implicate Order**
By Paavo T. I. Pylkkänen

**Particle Metaphysics**
A Critical Account of Subatomic Reality
By Brigitte Falkenburg

**The Physical Basis of the Direction of Time**
By H. Dieter Zeh

**Asymmetry: The Foundation of Information**
By Scott J. Muller

**Decoherence and the Quantum-To-Classical Transition**
By Maximilian A. Schlosshauer

**The Nonlinear Universe**
Chaos, Emergence, Life
By Alwyn C. Scott

**Quantum Superposition**
Counterintuitive Consequences of Coherence, Entanglement, and Interference
By Mark P. Silverman

**Symmetry Rules**
How Science and Nature are Founded on Symmetry
By Joseph Rosen

**Mind, Matter and Quantum Mechanics**
By Henry P. Stapp

**Entanglement, Information, and the Interpretation of Quantum Mechanics**
By Gregg Jaeger

**Relativity and the Nature of Spacetime**
By Vesselin Petkov

**The Biological Evolution of Religious Mind and Behavior**
Ed. by Eckart Voland and Wulf Schiefenhövel

**Homo Novus-A Human without Illusions**
Ed. by Ulrich J. Frey, Charlotte Störmer and Kai P. Willführ

**Brain-Computer Interfaces**
Revolutionizing Human-Computer Interaction
Ed. by Bernhard Graimann, Brendan Allison and Gert Pfurtscheller

**Extreme States of Matter**
On Earth and in the Cosmos
By Vladimir E. Fortov

**The Dual Nature of Life**
Interplay of the Individual and the Genome
By Gennadiy Zhegunov

**Natural Fabrications**
Science, Emergence and Consciousness
By William Seager

**Ultimate Horizons**
Probing the Limits of the Universe
By Helmut Satz

**Physics, Nature and Society**
A Guide to Order and Complexity in Our World
By Joaquín Marro

**Extraterrestrial Altruism**
Evolution and Ethics in the Cosmos
Ed. by Douglas A. Vakoch

**The Beginning and the End**
The Meaning of Life in a Cosmological Perspective
By Clément Vidal

**A Brief History of String Theory**
From Dual Models to M-Theory
By Dean Rickles

**Singularity Hypotheses**
A Scientific and Philosophical Assessment
Ed. by Amnon H. Eden, James H. Moor, Johnny H. Søraker and Eric Steinhart

**Why More Is Different**
Philosophical Issues in Condensed Matter Physics and Complex Systems
Ed. by Brigitte Falkenburg and Margaret Morrison

**Questioning the Foundations of Physics**
Which of Our Fundamental Assumptions Are Wrong?
Ed. by Anthony Aguirre, Brendan Foster and Zeeya Merali

**It From Bit or Bit From It?**
On Physics and Information
Ed. by Anthony Aguirre, Brendan Foster and Zeeya Merali

**How Should Humanity Steer the Future?**
Ed. by Anthony Aguirre, Brendan Foster and Zeeya Merali

**Trick or Truth?**
The Mysterious Connection Between Physics and Mathematics
Ed. by Anthony Aguirre, Brendan Foster and Zeeya Merali

**The Challenge of Chance**
A Multidisciplinary Approach from Science and the Humanities
Ed. by Klaas Landsman, Ellen van Wolde

**Quantum [Un]Speakables II**
Half a Century of Bell's Theorem
Ed. by Reinhold Bertlmann, Anton Zeilinger

**Energy, Complexity and Wealth Maximization**
Ed. by Robert Ayres

**Ancestors, Territoriality and Gods**
A Natural History of Religion
By Ina Wunn, Davina Grojnowski

**Space,Time and the Limits of Human Understanding**
Ed. by Shyam Wuppuluri, Giancarlo Ghirardi

**Information and Interaction**
Eddington, Wheeler, and the Limits of Knowledge
Ed. by Ian T. Durham, Dean Rickles

**The Technological Singularity**
Managing the Journey
Ed. by V. Callaghan, J. Miller, R. Yampolskiy, S. Armstrong

**How Can Physics Underlie the Mind?**
Top-Down Causation in the Human Context
By George Ellis

**The Unknown as an Engine for Science**
An Essay on the Definite and the Indefinite
Hans J. Pirner

**CHIPS 2020 Vol. 2**
New Vistas in Nanoelectronics
Ed. by Bernd Hoefflinger

**Life-As a Matter of Fat**
Lipids in a Membrane Biophysics Perspective
Ole G. Mouritsen, Luis A. Bagatolli

**The Quantum World**
Philosophical Debates on Quantum Physics
Ed. by Bernard D'Espagnat, Hervé Zwirn

**The Seneca Effect**
Why Growth is Slow but Collapse is Rapid
By Ugo Bardi

**Chemical Complexity**
Self-Organization Processes in Molecular Systems
By Alexander S. Mikhailov, Gerhard Ertl

**The Essential Tension**
Competition, Cooperation and Multilevel Selection in Evolution
By Sonya Bahar

Printed in the United States
By Bookmasters